Fantastic
Ferrocement

for

Practical, Permanent Elven Architecture, Follies, Fairy Gardens and other Virtuous Ventures

Peter Harris

1st printed edition 31 July 2006
revised, extended and updated July 2012

Published by Eutopia Press

P.O. Box 37, Kaiwaka
Northland 0542
New Zealand
Ph 09 4312 178
www.eutopia.co.nz
www.fantasticferrocement.com
email: peter@eutopia.co.nz

Table of Contents

4

Introduction

This book is mostly about the joy of ferrocement as a medium for the creation of durable beauty. The joy of having a concrete means to realise beautiful dreams that would otherwise be impossible without frightening amounts of money. Dreams of natural shapes, fantastical, whimsical, inspirational, sublime. Fountains, domes, follies, grottos, garden borders, birdbaths, ponds, pots, sculptures, steps, bridges, boats, towers....

The joy of ferrocement, like all joys, takes a little knowledge, provided here, but mostly just the courage to actually, physically try it. Your very first try can be useable, beautiful and above all a bridge to all the other possibilities that will suddenly open up to you when you realise how easy it is.

I hope you will use this book to build your bridge from the irrational 'I can't, because I never have,' to the triumphant 'I can, because I just did!'

Just do it' is a very wise saying, cutting through so much junk thought! There are so many skills I thought for years would be too hard to try, until I finally just tried them. Then they suddenly seemed easy—from then on it was just a matter of practice. This is a beginner's progress scale for the skill of ferrocement:

1. Buy this book. Congratulations! You're halfway there. (Plus, you have helped fund the building of Dreamspace, the ferrocement fairyland of inspiration dedicated to Beauty, Truth, Love, and Freedom!)

2. Buy the materials and a few tools if you don't already have them; you're 80% there.

3. Make a shape with the wire; you're 90% there.

4. Mix your plaster and plaster it onto your shape. Fantastic! You're 99% there.

Now you know how to do it, you'll be able to do it again any time. It's like riding a bike, and you'll get better at it every time, 100% guaranteed!

1 Origins

The first recorded use of ferrocement was by a country gentleman, Jean-Louis Lambot, who built a little ferrocement rowing boat in 1848. In a book I read in 1969 there was a photo of this pretty little boat—it was still in use on a lake. Lambot patented the method and planned bridges and other structures.

In 1849 a gardener, apparently working independently, used ferrocement to build flowerpots and, later on, garden furniture.

In the sixties and seventies there were a lot of ferrocement boats built, some very good ones professionally, in New Zealand. There was one plasterer in Whangarei who tells me his firm plastered over 400 boats! But there were many not so well done, and these gave the method a bad name. The main culprit was air pockets leading to rust and cracking of the hull. And the lighter, quicker-to-build fibreglass took over. It is a pity, as the chemicals used in fibreglass are not nice.

My own experiences: from ferrocement submarines and showers to Dreamspace and Café Eutopia

When I was a boy I read about ferrocement keelers and dreamed of building one. Then I forgot the dream until I went diving one day with a friend. I was entranced by the beauty of the undersea world, while nearly drowning through the snorkel, so I thought: 'Why not build a little ferrocement submarine?' Partly with this dream in mind, I left school and began building it in my parent's garage, but lack of money and practical knowledge, added to the distractions of being a teenager seeking the Meaning of Life, the Universe and Everything, meant that the rusting skeleton of the ferrocement dream was abandoned.

Submarines have to meet some very stringent requirements, which lie outside the scope of this book. Nearly everything else is a push-over in comparison. Still, it was many years before I got back into ferrocement. This time I was studying philosophy in Auckland, and we had just bought an old bungalow. It had no shower, and we were strapped for cash, so I decided to build one, and ferrocement sprang to mind. Despite Raewyn's misgivings, I went ahead. The results were quick and permanent. It was a pity I didn't bother to get some reinforcing rod—the chicken wire was a bit saggy, so we ended up with a wavy-walled shower. But it was good for singing in, and very, very strong. We found that we had built not only a comforting shower but also a reassuring earthquake shelter.

Then there was the think tank, my philosophy study for deep thought and escape from dogs, cats and young children. It was a seven foot geodesic sphere made of plywood triangles stapled together and sealed. It was great, but kept leaking until I plastered it and made a ferrocement vent on top, like a ship's ventilation funnel. The layer of concrete, though only about 2 cm thick, cut the noise right down and kept the rain out. The sphere has since been moved with us (by hiab truck), several times, and is now nearly fourteen, and sits outside Café Eutopia. It has been a jewellery workshop, an icon-sphere, an office for our recycled tyre pot business, then later the café office.

After two spheres, I decided the most ergonomic shape that still had the virtues of the sphere was the beehive—a hemisphere sitting on a vertical wall. And in the back garden behind a drystone wall overgrown with roses, the fairy garden we called it, I built, on a very tight shoestring, a seven-foot dome of old brick with a fibre- reinforced cement for the domed roof.

This led to the next phase, when we had moved out of Auckland, the philosophy studies over. We were sick of the infill housing and Raewyn had run out of garden space and yearned for an orchard as well, and a view of the sea, and a place for our children to bring our grandchildren into the world.

Thanks to my next project, it was to be years before we made this happen. Café Eutopia and Dreamspace (see vision-illustration) was, and is, a dream which grew in the making. The vision is to create a Camelot-like walled space of inspiring beauty and organic form, with a café-restaurant with domes, a fountain, a round table, a tower and a labyrinth, as well as rooms in the cloister walls for backpackers or a bazaar, and terraced herb gardens.

It all started in the 'little town of lights,' Kaiwaka, Northland, on State Highway One, just over the bridge, when I saw a glowing triangle of long green Kikuyu grass. We were visiting to look at a peninsula on the harbour which was later to become Otamatea Ecovillage. We didn't buy into the Ecovillage in the end, but the triangle of glowing green Kikuyu came up for sale some time later, and we bought it. I was compelled by a vision close to falling in love; Raewyn was compelled by a love for me close to martyrdom.

7

Our youngest daughter Xanthe said the field was like the Field of Dreams in the film by that name, and that is what we called it, not knowing what exactly it was for, but trusting that something beautiful would happen. We scythed the Kikuyu and Raewyn planted herbs and flowers, and I experimented with turning old tyres inside out to make big garden pots to keep the kikuyu at bay, When I painted the pots, they suddenly became a hot seller.

So Entyrely Recycled was born, yet another big stepping-stone and detour for me when I insisted, against Raewyn's better judgement, to move it to Auckland. There I lived a 'parallel life' with Raewyn, and Entyrely Recycled struggled with overheads, imports, bookwork and freight costs as I struggled to perfect the technique for converting the mountain of old tyres into beautiful painted pots. When I finally sold the business (at a loss) and came back to Raewyn and the Field of Dreams, a wiser man, I nevertheless knew that the original dream of domes still needed to be acted upon. But there were many inner resistances to be overcome. Still, being in Auckland I had been available to write for my brother John at Greenstone Pictures, and we had embarked on some wonderful fantasy adventure stories.

Also, I had done a workshop for finding your life purpose, called Lifework, started by an enterprising antipodean fairy woman, and there I had realised I was indeed a 'wizard' (which I had known all along—as one always does, deep down) and that I had neglected the divine feminine, and needed to return to Raewyn and Kaiwaka and the main purpose of my life, though it was still veiled.

Then one day I was taking our eldest daughter, Anna, back to med school. We stopped at the Dome café (which is not a dome, but overlooks Dome Valley). She said, 'I've always wanted to have a little café one day,' and suddenly a light went off. I saw that before making domes for sale, as I had been planning, but was stopped by my perfectionism, I could build a little domed cafe, on the Field of Dreams. And she said, 'If you build it, I will come,' or words to that effect.

So, applying the decisive 'just do it' approach which we had just learned at the Landmark 'Forum', combined with a mystical faith in the universal process, we declared our intention to not just dream this idea, but do it. I vowed to try to build the café dome for her before she came up again at the end of the term. I began, and she took the year off med school. When she arrived back at the end of term I hadn't finished it; only the foundations had been laid and a building consent got, but she got stuck in and helped. By now the plan was for five domes, one for the kitchen and four for customers to sit in.

We didn't know much about building, but we began to learn, the hard way. Our main obstacle apart from lack of money was fear of the unknown, such as council red tape (a greatly exaggerated fear, it turned out), my perfectionism, and my shyness about doing such crazy stuff in full view of everyone in town and all the vehicles streaming noisily past the little field. It was a year of idealism, growth of the vision, agony and ecstasy. Raewyn was scared of what we were doing in full view of all the world, and imagined a ruin with the Kikuyu growing over it, and us bankrupt.

But several bank loans later, it was ready to open as a little dream café-in-progress. We wanted it to be organic, which took a lot, but we did it and Café Eutopia was born. Eutopia means Good Place in Greek, as opposed to Utopia which means literally, 'Not a place,' somewhere fictional, a place assumed too perfect to exist in real life. We had a magical 'pre-opening' party, and some good friends we had met spoke of how the vision had found a place in their hearts and inspired them to help. Then, when we realised the full extent of the project and decided to accept it would be a work in progress for many moons to come, we held the main opening, and the good Mayor of Kaipara came and declared café Eutopia open, and there was dancing and celebration under the temporary tent where the Chartres labyrinth will one day be built.

The whole Big Vision, which I called Dreamspace, continues, the blueprint written on our minds and hearts, and some of it on paper. And batch by batch we plaster and carve and paint, as money and time permit. By acting on my deepest impulses and values, 'The jagged pieces of my life have come together to form a complete, mystical whole' (from Hook, the movie). I now see that whole as a temple to 'Beauty, Truth, Love and Freedom.' And as a sculpture of the Universe, and an embodiment of the marriage of Heaven and Earth, the marriage of opposites.

After Anna returned to med school, Raewyn nobly left Koanga Gardens where she had been working and took over the struggling little café, and I continued writing for my brother to make ends meet, and to convey something of my eutopian vision of how life can be. And if one day I write a best-seller, the rest of Dreamspace will be built, a fantasy in ferrocement, glass and wood, founded, like all Eutopias and Camelots and 'dreaming spires,' on visions written and woven over the years on the enchanted loom of the human mind, built on the good earth with sacrifice, love, and long hard labours.

Esoteric ferrocement

The symbolism of ferrocement is interesting. The uncrushable but brittle married to the bendable but unbreakable. A marriage of opposites. I had these impressions as we built Dreamspace: the wire framework is like the bones of the body; the chicken wire the muscle layers, and the plaster is the skin. Or, the wire framework which is woven together is like a spider's web, beautiful in its own right, ethereal, the wind whistling through it, but strong and supportive. Then there is a transformation, and that beauty is lost, but another is born: the beauty of the solid enclosing form. Then even that is transformed when it is carved and painted, losing its starkly beautiful bone-like look and gaining whatever colours our aesthetic sense dictates.

Esoteric, frustrating ferrocement

And the journey of building in ferrocement is a good exercise for would-be creators. It is a discipline of actually grounding a pure perfect beautiful heavenly idea in those two very stubborn and earthly materials, steel and concrete, hard and heavy, recalcitrant and messy. Sometimes as Anna and I laboured on the domes and arches of Dreamspace the steel got very rusty and poky and springy, and the concrete got very wet and cold, messy and abrasive, getting into all the cuts the steel had made, ruining clothes, and dropping off the mesh or sliding off trowels into gumboots, splattering tools and ladders with mud that turned surprisingly quickly to stuck-on stone. And just when we thought we were finished, we found we weren't; there was always another area to do or another coat to put on. Then the daylight always seemed to run out and we'd be cleaning up in the dark.

The need to keep the plastered areas wet for seven days was another trial of patience. Our lovely work hidden, covered in ugly sacking and dripping with water from the hose, day after day. And if this was skimped, the plaster could dry too quickly and become less strong, and its future for all time be less secure.

In other words, doing ferrocement properly is like doing anything of a permanent nature properly. It takes knowledge, courage, patience, preparation, hard work, and faith in the value of the end result. So, what else is new?

2 What is possible?

Given the above heroic human qualities in large quantities, the sky's the limit. My vision is of a kind of renaissance culture—we are not far off it in many respects, here in New Zealand—which reverently yet exuberantly combines the best of old and new. Like all great cultures, it will need an architecture for every level. For homes, stone and earth and durable natural wood are good, but for sculpture and inspirational buildings in organic forms, ferrocement is perfect. It is a fraction of the price of stone, but plastered with an extra layer and carved it is a durable creative medium and can be safely formed into shapes impossible with stone, though the wonderful gothic cathedrals show what can be done even with nothing but stone and wood. But try getting a building consent for a gothic cathedral built of unreinforced stone! And getting a banker to finance it!

So far, over three years off and on, we have built five ferrocement domes joined with arches around a courtyard with a ferrocement fountain carved as a blue lotus, a part of the cloister wall, with pillars carved with two fairy-gnomes, and a moongate with a 6-metre guardian gull over it. And two ferrocement tables, one with a spiral support, the other carved with a picture of a man and woman, and the caption 'Live hand in hand, and together we'll stand, on the threshold of a dream.' (from the Moody Blues song). Most of the place still hasn't had its final coat of plaster which is to be carved then stained or painted. But good things take time.

Our next big push is a kitchen extension with a bar opening onto the Labyrinth circle, all in ferrocement. The engineers who did the calculations were intrigued but happy with the practicality of our plans. And if engineers are happy, and you take care with the actual construction, in our experience the building inspectors will be happy too. After all, the Sky Tower in Auckland is built of concrete! Our aims are so much more modest.

Is there a new type of ferrocement just around the corner? Could be, where the mesh is woven in situ, for example. Tiny robot weavers, for example, weaving the framework to a shape designed onscreen? It would still need bigger reinforcing, but maybe this could also be woven, like steel cable?

11

FIGURE 1 An early blueprint for Eutopia – lots more to do before it looks like this!

Figure 1A A later (2007?) version. Even if something like this never materialises it is worth the envisioning, and informs the smaller things we do manage to do....

3 The Concept, Costs and Benefits of ferrocement

The concept of ferrocement

The idea is that if you impregnate a fine enough network of steel with a thin layer of cement plaster (that is, concrete without the gravel, just sand and cement), you get a very strong, flexible 'membrane' with the properties of both steel (Latin 'ferro-') and concrete; hence the name ferrocement. And because steel and concrete both expand and contract at the same rate, it doesn't crack easily with extremes of hot and cold weather; also the rich cement layer 'passivates' the steel, stopping it from rusting as long as there are no air pockets or cracks.

For small objects made of ferrocement, the steel network can be made of just the fine strands of reinforcing, but for bigger objects the idea is to create a fractal network of reinforcing, i.e., to have a coarse network of thick reinforcing, then a finer network of thinner reinforcing over it, then finer still, down to the finest.

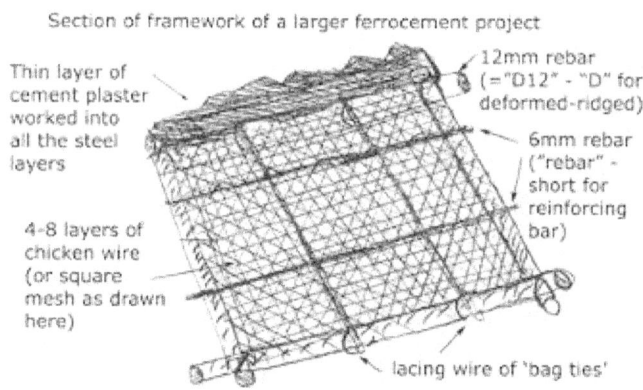

FIGURE 2 Section of framework for a medium-sized ferrocement object

Practical ferrocement technique - overview

There are many possible ways to achieve this steel-rich cement membrane, but typically, two to eight layers of standard galvanised 'chicken wire' (12mm mesh size) from a roll about one metre wide are tightly tied with thin wire onto both sides of a framework of reinforcing rods, then a strong cement plaster is trowelled or hand-smeared into it, in one or two coats. The mix is usually about 2:1 or 3:1 i.e., 2 or 3 containers of sand to one of cement, well mixed with water with a little plasticiser added for ease of working. The water is added until the mix is like porridge, not too thick or too thin. The right mix will be easy to push into the mesh layers, and ooze through the other side of the mesh like toothpaste, without slopping right out like runny porridge would. It is hard to describe thickness, but you will soon get the feel of it. Experiment with small batches.

Variations: expanded steel mesh can be used, but it is not as good— the steel is too thick and the plaster can crack and peel off it. Welded square mesh, 12 mm mesh, can be used, but it is more expensive and can't be wrapped over curves that go both ways (e.g. a ball shape) like chicken wire can, and even on flat curves I have found it tends to buckle in and out when tied to the reinforcing framework.

What ferrocement is best for - a cost-benefit analysis of the method

The materials, thanks to modern knowledge, are really very cheap, for such a permanent strong building material. But if you are looking for the cheapest, this isn't it. Ferrocement is nowhere near as cheap or quick as four by twos and play and corrugated iron, for example. But if you are willing to do it yourself, and are determined to have curvaceous forms in a permanent material, ferrocement is for you, in fact there is hardly any (building-permitable) alternative.

Costing (NZ dollars - as at 2007) per square metre for a reasonably big object that needs 12mm reinforcing rods as well as 6mm:

12mm (D12) mild reinforcing rod crisscrossed to make a 500mm 'mesh': 4 m at $1.16 = $4.64

6mm mild steel rod crisscrossed to make a 200mm mesh:25m at $0.66 = $16.65

12mm mesh galvanised chicken wire 900mm wide, four layers = 4.44 m at $2.40 = $10.65

Lacing wire to tie the chicken wire on: say 7 m at $0.07= $0. 49

Total steel cost: $31.94

Plaster at average 20mm thick = 0.02 cubic metres, less steel = 0.0138

Say 0.6 is Sand: at $60/cubic metre = 50 cents

Say 0.4 is Cement at $10/ bag = $500/cubic metre =$2.76

Plus plasticiser at $15/litre 5mls = say 10 cents

Total plaster cost = $3.36

Total material cost for 20mm thick ferrocement panel: $35.30/square metre.

Or, if it needs only the chicken wire (for a small item): $14.01/square metre

This may seem cheap. But the labour content is high. For a shape like a 2.5 metre dome of say 30 square metres of wall and roof, not counting detailing like gutters, doors and windows, just the flat surfaces:

12mm bars bent and tied in place: 4 hrs

6mm rod bent and tied in place: 16hrs chicken wire cut and laced on: 24 hrs plastered inside and out, first coat: 16 hrs plastered, second coat if needed: 12 hrs final coat sponged/rubbed/scraped 6 hrs

Total 78 hrs, divided by 30 square metres = 2.6 hrs/square metre, or say $52 at a skilled worker's rate of $20/hr (to do it at this speed and still do a good job, you would have to be skilled. Otherwise, triple it!).

And counting detailing and 'fiddly bits,' I would double it to say 5.2 hours/square metre. Plus all the designing and trial and error for one-off structures, say another 0.5 hours. And the second coat and brushing/scraping of the final coat, say 0.25 hrs. And the watering and checking, say 0.125

Grand Total labour 6.075 hrs per square metre. Round it off to 6hrs/ square metre = $120/square metre.

So, for total materials and labour for a single skin of ferrocement (if it is a building, it would need insulation and an inside layer as well) you are looking at about $35.30 materials plus $120 labour = $155.30. Adding 5% for waste and costs like rubber gloves etc, say $163/square metre.

So (Updating to 2012 costs, adding 50% since steel and cement have gone up a lot!)the completed shell of a 2.5 metre diameter dome (floor area 4.91 square metres) with two window openings and a door opening, without floor pad, and before painting, carving, etc, and installing windows and door, would cost $7335, or $1493/square metre. So, there are much cheaper ways to build a garden shed!

But if you are willing to do it all yourself, sacrificing the time and energy, the actual money outlay is much lower: $1588 for a 2.5 metre dome, or $323/square metre of floor.

And if you are sculpting small shapes or making large garden pots, for example, the cost is low compared to stone, the result is probably stronger, and you can make shapes that would be impossible in stone.

Other pros and cons

Possible electromagnetic disturbances

Some green builders say living in a 'Faraday cage' i.e. anything made of steel, is bad for us. I don't know enough to comment, but most ferrocement things are not made to be continuously lived in anyway. And if it is a problem for ferrocement, it is also a problem for a large number of people living and working in steel buildings, and driving in cars etc. If there is a danger, it is a low-level one, it seems. And just stepping out our front door is dangerous…

Environmental concerns

These are mainly about the amount of energy used in producing steel and cement, compared to other materials like wood. But we have to remember ferrocement is permanent so the cost is (almost) one-off. And there is actually very little material in a ferrocement structure, compared to, say, a cast concrete one.

And finally, it is not possible to create all these beautiful forms any other way, not permanently and with minimal chemical or heavy metal pollution. I think the high energy input, if the energy is hydro and not nuclear, is definitely worth it, to make human environments more beautiful and inspiring.

4 The Nitty-gritty:How to make fantastic ferrocement

If the last section has not put you off, you are obviously a seeker of lasting beauty and are willing to pay the price. Read on, kindred spirit!

A. Tools And Gear You'll Need

For Steel Framework

1. Hacksaw or Bolt cutters or angle grinder or metal cut-off saw (for cutting the reinforcing rods)

2. Pliers (for general twisting and cutting of lacing wire and chicken wire)

3. Big old pair of scissors, or tin snips or—quickest by far: an angle grinder with metal cut-off wheel (for cutting chicken wire)

And, if you are doing a lot of ferrocement:

4. Angle grinder is now a must!

5. Several sizes of concrete ring (or tanks or similar) to make large circular bends in reinforcing rods

6. Bender for sharper bends to the reinforcing

7. Welder or access to one

8. Extra-large adjustable pliers (good for bending 6mm rod around thicker reinforcing, etc—though it can all be done with the grooved-pipe bender—see below)

Tools you will probably need to custom-make:

9. Hook for twisting bag-ties (to go into a variable speed power drill, preferably cordless)

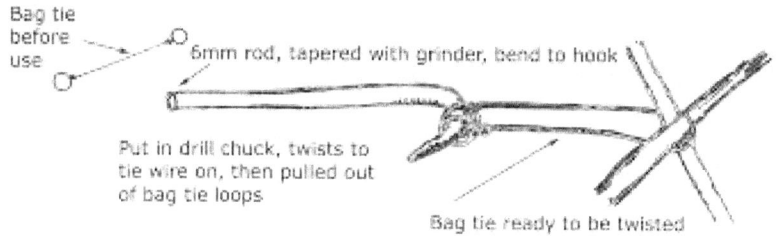

FIGURE 3 Hook for twisting bag ties (to go into variable speed power drill, preferably cordless)

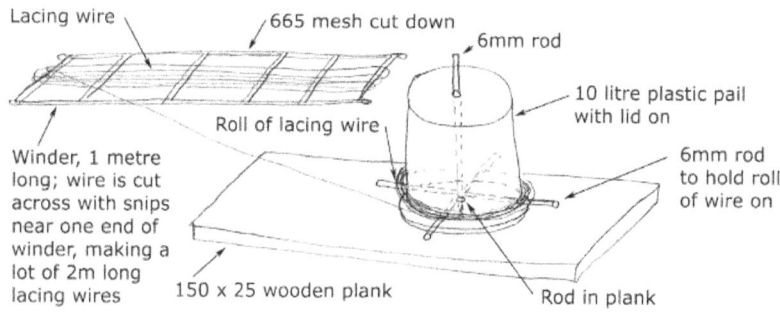

FIGURE 4 Lacing-wire winder for cutting into lengths

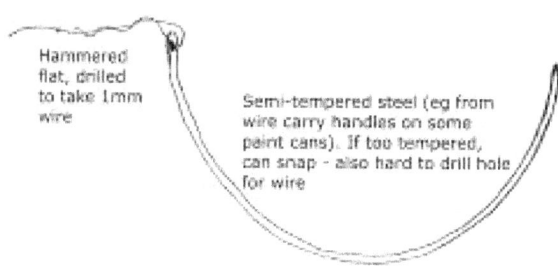

FIGURE 5 Semicircular lacing needle

19

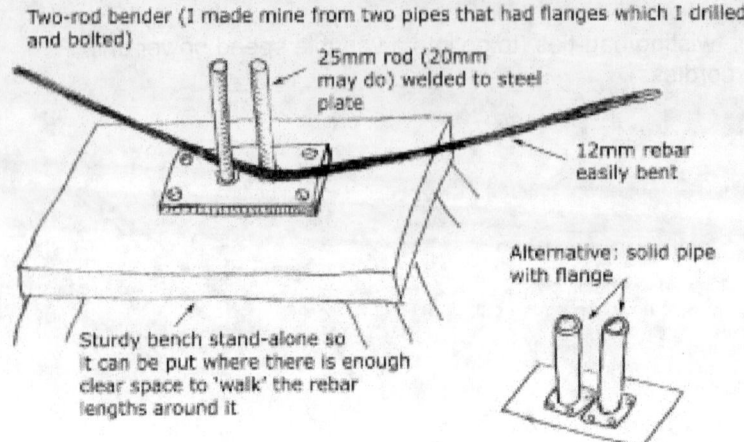

Two-rod bender (I made mine from two pipes that had flanges which I drilled and bolted)

25mm rod (20mm may do) welded to steel plate

12mm rebar easily bent

Alternative: solid pipe with flange

Sturdy bench stand-alone so it can be put where there is enough clear space to 'walk' the rebar lengths around it

FIGURE 6 Two-rod bender

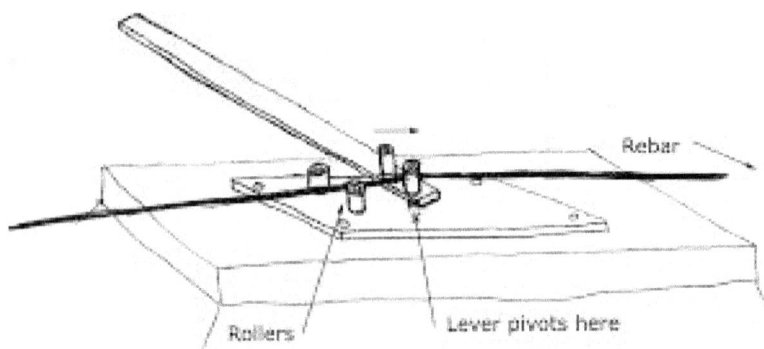

Rebar

Rollers

Lever pivots here

FIGURE 7 Commercially made bender

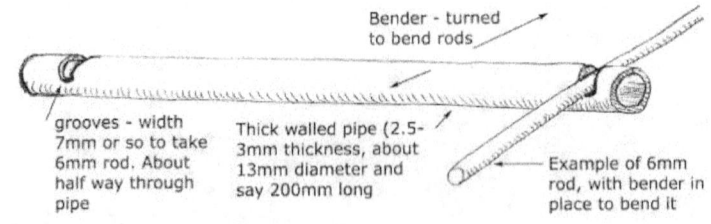

Bender - turned to bend rods

grooves - width 7mm or so to take 6mm rod. About half way through pipe

Thick walled pipe (2.5-3mm thickness, about 13mm diameter and say 200mm long

Example of 6mm rod, with bender in place to bend it

FIGURE 8 Hand-held grooved pipe benders for bending thin reinforcing (my own design—or has it been done before?)

2 short pieces of 12mm rod welded to bent-over handle piece to make 2 'prongs'

FIGURE 9 Hand-held bender for thicker reinforcing (got this from the internet)

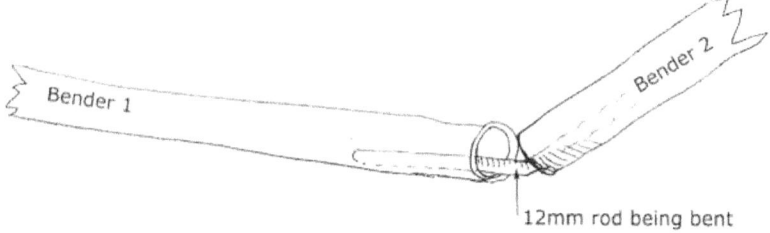

Bender 1

Bender 2

12mm rod being bent

FIGURE 10 Pair of 25—30mm diameter pipes,1-2 metres long for bending thicker reinforcing

For Plastering

1. Wheelbarrow (for mixing/carting)

2. Plastic buckets (for measuring sand and cement accurately, and carrying plaster to the work)

3. Shovel

4. Rubber gloves

5. Round-ended trowel/s

6. Water hose

7. Ladder of suitable size if project is higher than 1.5 metres or so. A sturdy old chair is handy too, for lower parts, and to put things on.

FIG 11 Round-ended trowel

And, if you're doing a lot of ferrocement:

8. Power concrete mixer, ideally a plaster mixer

9. Several sizes and shapes of trowel

10. Scratcher: handheld 'rake' for scratching lines into first coat of plaster while wet, for better sticking of second coat.

11. Hawk: Square board with handle to hold plaster and refill trowel from

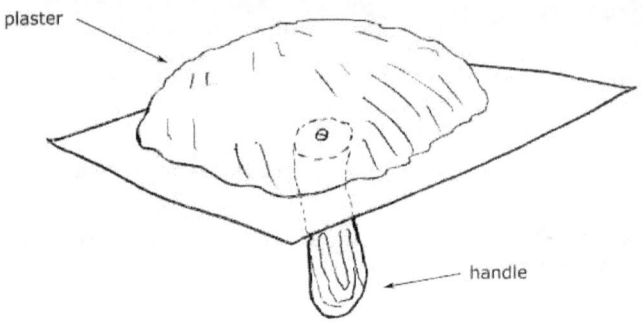

FIGURE 12 Hawk for holding plaster

For Polystyrene Work

1. Cartridge gun and water-based construction adhesive that doesn't dissolve the polystyrene.

2. Sharp knife or snap-off cutter, or ideally an electric hot-blade knife. This is great for shaping and carving polystyrene as well as just cutting it.

3. Special fibreglass cloth and tape from a plasterer's supply shop. It is coated with plastic to stop the cement from dissolving the glass fibres (the alkalis in cement react with glass and weaken it, so all glass fibre reinforcing used with cement products has to be coated. You can't use the ordinary glass fibre used with epoxy resin to make 'fibreglass'.

22

4. (Expanded) Polystyrene sheets—20mm is good, or 25mm, and 30mm or even 40mm for bench tops etc. Also, you can get blocks of it and carve it, then plaster and fine carve. But this is another technique, and I think isn't as strong and satisfying as real ferrocement with a steel framework. And the polystyrene smells when you use the hot blade, and produces lots of light, nuisance fragments when you sand or file or saw it. Raewyn hates it, partly because hens can eat it, thinking it is grain, and it apparently floats in their digestive system and bloats them so they don't eat. On the positive side, it is 98% air, and doesn't rot, so for insulation, as long as rats can't get at it, or heat melt it, it is great. And it isn't as toxic as Urethane foam, which gives off really nasty fumes when it burns. According to the official polystyrene brochures, put out admittedly by the industry, polystyrene smoke is no more toxic than wood smoke. But it certainly doesn't smell as nice! Nowadays it is sold with flame retardant in it, so it will not burn, only melt, unless a flame is held to it and kept there.

For Carving

1. Tungsten scraper, such as the one made by Trevor Lindsey (Linbide).

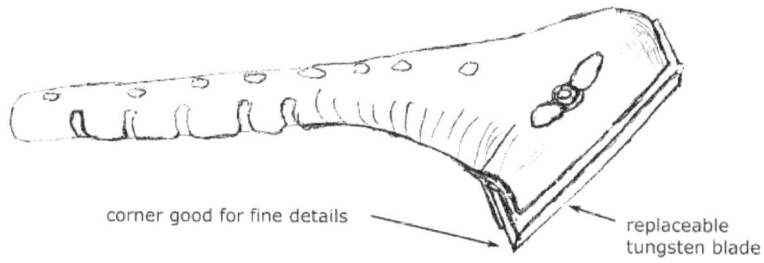

corner good for fine details

replaceable tungsten blade

FIGURE 13 Tungsten (paint) scraper

This is all you need for just about all carving and scraping of soft or even fully cured plaster. The anglegrinder can also be used on harder plaster, but it makes a lot of dust, and you would need to wear a good mask.

B Materials For Ferrocement

Reinforcing

1. Thick lacing wire, low-tensile mild steel, 'black,' i.e. ungalvanised, about the thickness of fencing wire. Comes in rolls. Cut with pliers, angle grinder or bolt cutters. For:

- zigzagging between thicker reinforcing to make a wide 'rib,'

- reinforcing in smaller objects, especially free-form sculptures, where it is handy as it is so easily bent

- winding into coils or 'springs' to hang from walls and plaster for solid pillars

2. Square mesh reinforcing sheets '665' is 6 mm rods welded into a sheet 2 by 4.7metres with a 150mm square mesh. Good for larger cylinder shapes or flat panels. Cut with bolt cutters or angle grinder.

3. 6mm reinforcing rod (R6), mild steel, comes in 6metre lengths. Cut with angle grinder or bolt cutters or hacksaw.

4. 10 and 12mm 'deformed' reinforcing rod (D10, D12) i.e. ridged for better grip in the concrete. Mild steel, comes in 6 metre lengths. Cut with angle grinder, or extra-large bolt cutters or hacksaw.

5. Chicken wire, 12 mm mesh, comes in 50 metre rolls, about 900 mm wide. For the final reinforcing which holds

the plaster. Cut with angle grinder or tin snips or large scissors.

(Welded mesh, 12 mm mesh, comes in 30 metre rolls, about 1metre wide. For the final reinforcing which holds the plaster. Cut with angle grinder or snips. Can be good for flat panels or cylinders. Stronger than chicken wire, but not my favourite as harder to work, less flexible and dearer).

Fastening

1. Thin lacing wire, usually galvanised, preferably 1 mm thick. Comes in rolls. Cut with pliers, tin snips. Used in 2metre lengths with lacing needle to lace chicken wire layers together.

2. Bag ties, short lengths of wire with loops at each end for twisting around anything (usually sacks, but reinforcing in our case). Used with bag tie hook in drill to tie two or more reinforcing rods together.

3. Thick lacing wire, cut to suit for tying thicker reinforcing.
Also used for reinforcing in its own right.

For the Plastering

Sand
Should be 'sharp' washed river sand ideally, not salty or muddy or rounded. Should have a good mixture of grain sizes from coarse to fine, and as little shell as possible. Ask for Plastering sand. Keep clear of gravel which is a pain when plastering—each bit has to be flicked out! Store under cover to avoid rain, wind, cat droppings and dead leaves etc.

Cement

Ordinary bagged Portland cement. Not lumpy from being stored too long or getting damp.

When you are confident, get the rapid hardening cement used e.g. by water tank manufacturers. Handy to avoid slow sagging when building up thick layers for carving. And to lessen the chance of rain coming before the plaster is dry enough to cope.

Plasticiser

Ask for the normal liquid plasticiser used by plasterers and bricklayers (NOT 'superplasticiser' which is used in concrete to reduce the water content but makes plastering a nightmare, as it changes consistency as you work it).

Adhesive

This is used as a glue to brush on between coats to help them stick, if the undercoat of plaster is not rough enough to give a good 'key.' As an additive to make the plaster more impervious to water, more plastic to work, and stick better any surface. It is expensive, but worth using when needed.

Paint

After the plaster has fully cured, say 2 or 3 weeks, any good acrylic can be used on it. There are special paints for coating uncured concrete, and special floor and bench coatings for heavy wear. Also there are roof paints—acrylic is ok if there are no cracks. And glazes— varnishes—for coating delicate paint finishes such as wiped/ragged effects.

C Safety And Comfort

Safety

Wire and reinforcing rods can poke eyes, or if they are sticking up, impale. So care is needed. Also, reinforcing mesh can spring out dangerously if bent e.g. in a circle, and suddenly released, and the ends of the mesh, especially if bent up, can be deadly.

Angle grinders can cut, the disks if damaged can fly off, bits of disk or the metal dust can get into eyes, and the sparks can cause fires. So take care of the disks, wear eye protection, and keep flammables away from where you are grinding metal.

Electrical tools in general can cause shocks outdoors—always connect to an isolating transformer or RCD cut-out switch. Children especially need to be kept away from tools and materials. Cement dust and concrete dust are not good for eyes and lungs, so don't let dust blow about, and wear a mask if grinding concreteor plaster.

Concrete mixers can be dangerous—make sure the gears or belts are properly guarded so hair or hands can't get caught in them. And don't put hands into moving concrete bowl.

Ladders need to be in good order and placed securely, and moved so you don't over-reach.

Comfort and keeping clean

Nuisance cuts to hands from chicken wire can get irritated by the alkaline cement, and skin can get very dry and wrinkly after plastering with bare hands. Always wear gloves. I prefer latex gloves—they are very cheap by the boxful, and you still keep flexibility and feeling. They are great for hand-smoothing of plaster,

26

e.g. on edges. And to minimise cuts and pricks from chicken wire, try leather or thick rubber gloves such as electrician's high voltage gloves (we got some from Arthur's Emporium cheap. Surplus stores can be very useful!)

Clothing: when plastering, it's hard to beat gumboots, preferably with steel toes, and light, stretch-material wet-weather leggings. Waist down is where most of the wetting and spills happen. And above the waist, long-sleeved old tee-shirts are good, to keep the spots of plaster off skin. (otherwise, if it is too hot and you have bare arms, wash them off regularly). A hat is good, to keep sun off and any falling bits of wet plaster out of the hair. Tie long hair up.

Skin moisturisers are nice, especially before plastering, to keep the cement out of pores.

Drink plenty of water—from bottles so it stays clean.

Plan work conservatively so you don't run out of daylight!

Take regular breaks, so you don't get overtired and make mistakes - and be too stiff and tired the next day!

Always clean up meticulously in daylight or good floodlight—it is amazing how many places plaster can be hiding, to turn into rock if not washed off. And tools are much better smooth and clean. Have a good water supply and trigger hose fitting preferably. And don't let lots of plaster and sand wash into drains where it will harden and block them. It doesn't seem to harm garden soil or plants, though—in moderation.

Ask at safety and protective clothing shops for all this gear.

5 Step By Step General Instructions, Techniques And Tips

In later sections I will describe a variety of example projects to cover the main variations of size and shape. You can adapt the instructions for the one that is closest to what you want to make.

A: Planning

It is important to have a plan, but until you are familiar with the medium, it is hard to know exactly the best design for the details of the framework. Don't worry—you will learn by doing. You can modify a framework as you go along. Also, look at the example projects.

Things to bear in mind when designing in ferrocement:

- The ferrocement is normally thin—20mm or so—and so all openings should be rounded, given a lip, so that it doesn't look too thin. I like the method of solid-plastering pillars and branch forms around the openings to give a really solid look. This is done after the ferrocement shape is plastered, and means you don't have to do complicated reinforcing for it.
- The chicken wire, and then the plaster, adds to the thickness by about 12mm on either side of the reinforcing framework, maybe 20mm around openings. So, for example, make a window opening about 40mm larger in diameter at the framework stage.

- Ideally both sides of the object should be reachable for plastering, even if one side (e.g. the inside of a sculpture or a pot) is not normally seen, other wise the un-worked side will have some of the reinforcing exposed to the air and possible rusting as well as making the structure weaker. A way around this is to pour plaster or concrete into the inside of the object and fill it up ('grout' it). This makes the object heavy, and may be expensive and also require a grouting hole to be left at the top of the object, if it is too big to be turned upside down.

28

- The object will get surprisingly heavy if you put extra coats on for carving or to cover the wire if you didn't plaster evenly over it on the first coat. So, plan for lifting and moving challenges unless the object is small, or you build it in place. An advantage of ferrocement is that it can be rolled around without breaking in pieces, and fork lifted or craned. And it can be easily repaired if chipped.

For deciding how thick the reinforcing needs to be, consider:

- the reinforcing has to be firm enough to hold the shape while the plaster sets, supporting the weight of the wet plaster without sagging. And to hold your weight while plastering, if it is a big object (unless you go to the extra bother and cost of scaffolding). Plastering from ladders is possible, but even then you want to be able to lean the ladder against the framework.

- It has to be strong enough for all reasonable and sometimes unreasonable stresses and strains on the finished object (allowing a margin for such things as moving, rolling, forklifting, dropping).

- You will need less or more reinforcing depending on the shape—less if domed or arched, more if flat or square. The weakest point is any opening in the object, so, for objects with a lip or rim, like pots, the thickest reinforcing is needed around the rim. This can be either two or more
6mm rods laced together, or one 10 or 12mm, or even two of these for bigger things like domes. The other way to strengthen openings or rims is to make a wider rim, i.e. bend the lip over to make an 'L' section. It is best to use reinforcing rods for extra strength rather than extra layers of chicken wire, because it becomes very hard to get the plaster into more than about eight layers of chicken wire. And air 'voids' are bad because they weaken the structure and, worse, can allow rusting from the inside, which will eventually crack the plaster open as the rust expands.

Architects, Engineers and Red Tape

For bigger structures, consult an engineer, preferably one who knows ferrocement. And even before that, an architect, who will know of other alternative building methods which may be more economical, or suit you better. Or a combination of ferrocement and other materials.

And see your local council building inspector before doing final plans, and find out if you need a building consent. It should be easy enough, with such a strong material, and you will save a lot of stress if you do it legally! You may have to specify in your plans heavier reinforcing or foundations etc, than you would have done otherwise. Try and accept this - the bylaws do enshrine a lot of commonsense and wisdom from long experience of such dangers as earthquakes, floods, and also human error, negligence and deviousness. They are designed to protect everyone, including innocent children, bystanders, and those who may buy our creations.

Councils do differ in their attitudes, but my experience in the Kaipara is that they want to be helpful and allow freedom for people to be different—as long as it is safe. Innovators and protectors of the status quo need to work together, and they do to a great extent in our society. Naturally they are suspicious of each other, and don't necessarily like each other much. But speaking to the artists and inventors, on behalf of the protectors: red tape and officialdom are far worse in our minds than in reality if we check the rules out first, and plan within their limits. Many great ideas don't happen because the inventor assumes he or she will be opposed and throttled by 'narrow-minded bureaucrats' and 'red tape'. But ninety nine percent of the red tape is for stopping those who ignored the vital one percent, which was (mostly) only put up to keep the inexperienced from wandering off the edge of a cliff.

There is a cost to doing things differently—the rules can't cover all cases, all building methods, so instead they allow for experts to judge the unusual cases. Engineers, usually. And experts aren't cheap, and err on the side of caution. But if a thing is worth doing, it's worth doing legally. You might even make some new friends and supporters for your project where you least expected!

Drawing up plans

This is fun, and good to do thoroughly, as long as you don't get obsessive and bogged down! First do concept drawings before getting into construction details. It's your creation, your baby. Get clear what you really want it to look like.

Then look at practical details as above, how the reinforcing should go, where the openings—doors and windows, openings for plastering the inside—should be; how rainwater will run off it, etc. You may have to change the plan a bit, but at least you know what you want ideally.

30

Making a Model

I nearly forgot the other planning approach: 3-D modelling. I don't use CAD or similar on computers, though this could be great, but I make little plaster models of big projects. If you are at all confident in carving, I urge you to try making a plaster model. It is great fun, and can help convince the sceptics that your dream project is going to be real, and look good. Plaster of Paris is a wonderful material for carving models, cheap (especially if you get a 20kg bag) and easily cast in a plastic container etc. and easily carved. For odd-shaped projects you can make a cardboard mould to cast the rough shape—I have used strips of card or plastic cellotaped together to the shape and then taped down onto a smooth sheet, like a piece of plate glass.

I use either casting plaster or any plasterer's gypsum plaster, sometimes called hard plaster. It sets in 10 minutes or so, and can be carved even while still damp. If you want it dry and chalky, for painting, put it in the sun or even in the oven.

Chisels are okay for carving plaster, and hacksaws for cutting blocks—or even a wood saw though it clogs the blades a bit—though this is easily removed. I use an engraver's chisel I made for fine work, with a diamond-shaped point.

I use a scale of 1cm to 1 meter for bigger projects, 2-4cm to 1 metre for smaller. Usually I make a model only about 50 to 100mm long. Otherwise it takes too much carving and the model can take too much energy away from the project, or never get finished!

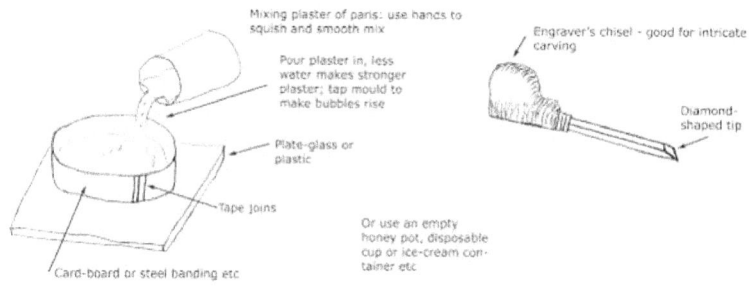

FIGURE 14 Model making ideas

Calculating materials needed

This is an art when the shape is free-form. Estimate and add a margin for error. It is less important if you know you can use the left- over materials, and if you can afford it!

For calculations, use approximations to the shape you want— cones, cylinders, circles, hemispheres, rectangles. And subtract the openings from the area.

For the chicken wire

Calculate the surface area, then multiply it by 1.1 to allow for overlap multiplied by the number of layers (usually four, giving 4.4), divided by the width of the chicken wire in metres (usually 0.9). This is hopefully the number of metres you will need.

For the reinforcing

Get the circumferences and lengths and widths of the various surfaces, then work out how many 6 metre rods, at the right spacing, you will need, for each thickness needed. Add a margin of at least
20% (multiply by 1.2) for offcuts, overlaps etc. The offcuts will end up being used, mostly, for extra bits.

For the lacing wire

It is surprising how much you use. About 7 metres per square metre of ferrocement.

B: Foundations

This section is only relevant if you are building a ferrocement building, or laying ferrocement on the ground, e.g. for a pond.

For the Dreamspace domes we used a normal concrete pad (ready mix from a truck) of about 100mm thickness, reinforced with '665' welded square mesh, with 'starters' of D12, 12mm reinforcing rod, to tie the wall framework to. See building regulations for your area (ask the building inspector for the relevant regulations), but you should be fine if you spaced L-shaped starters 600mm apart along the wall, overlapping the wall by 600mm. The concrete foundation floor can be plastered over later and carved into. Where the ferrocement walls were to go, we painted the concrete with a sealer to prevent 'wicking,' the drawing up of moisture from the ground into the walls.

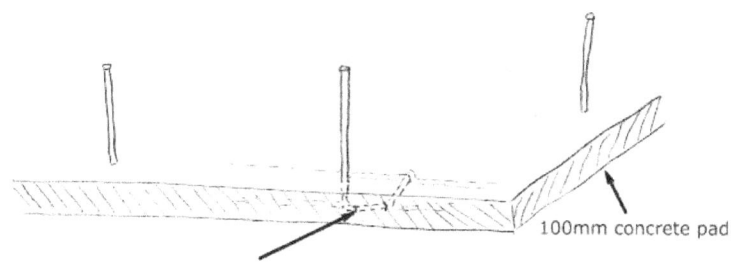

100mm concrete pad

D12 starter, bent to be wired to rebar in concrete and stick up about 600mm

FIGURE 15 'Starters' in a concrete pad

If you are laying ferrocement on the ground, it is a good idea to sprinkle a layer, say 10mm thick, of sand over subsoil if any, to reduce cracking. I have made steps in a bank by digging them out of the clay, carving them to size with a sharp spade, then laying four thicknesses of chicken wire down, bending it up the sides of the excavation as well, and plastering over it. After a second coat had dried overnight, it was ready to carve a paving-stone effect into. I made a winding ferrocement path the same way.

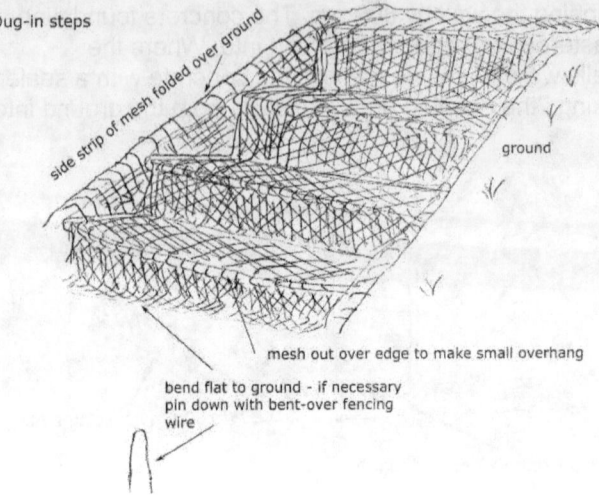

Dug-in steps

side strip of mesh folded over ground

ground

mesh out over edge to make small overhang

bend flat to ground - if necessary
pin down with bent-over fencing
wire

FIGURE 16 Ferrocement steps dug into a bank or slope

<u>C: Reinforcing</u>

This comes in 6 metre lengths that are often a bit rusty. This is
normal, and the cement 'passivates' the rust. What you don't want
is oily reinforcing, and if it is oily it should be left out in the weather
for a while or else water blasted with detergent. We have never had
to do this, though.

Now comes the challenging bit where you have to bend the rods (or
thick lacing wire for smaller items) to get the 3-D shape you want.
Start with the vertical rods, if possible. (e.g. in a pot or dome) Or the
ones that go along the longest side of the object (e.g. on a long
horizontal object like a dragon).

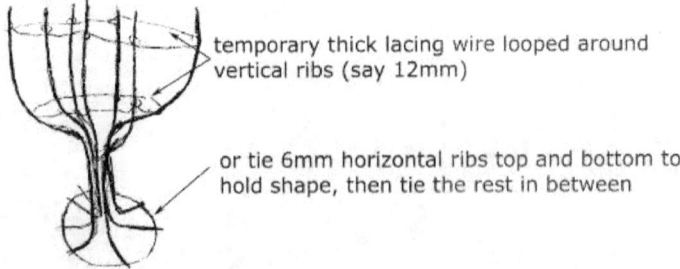

temporary thick lacing wire looped around
vertical ribs (say 12mm)

or tie 6mm horizontal ribs top and bottom to
hold shape, then tie the rest in between

FIGURE 17 The vertical ribs tied together

34

The thickest rods form the framework on which to bend and tie the thinner rods, so if there is a standard, profile, e.g., of a pot or dome, it pays to bend one piece carefully to shape, then use it as a master to get the others the same. It is worth spending some effort to get them right, as they are the basis for the whole shape. After that the thinner rods are easy.

After you have the number of thick rods you need for the vertical profile, you need to tie them in place and then make the horizontal rods (which I often make of thinner rod than the vertical), then tie them onto the vertical ones. It can be tricky to hold the first rods in place. Put them on a level floor or ground, and use helpers (if available) to hold them while you tie thick (2 to 4 mm) lacing wire between them for temporary positioning. Or use 6mm rod pre-bent to the horizontal profiles. And if you have a concrete pad with starters, tie the verticals to the starters. (Or, if you are using reinforcing mesh for walls, bend this around the starters and tie it in place, then tie the verticals to it).

All the first tying is quickest done with the bag-ties and hook turned by a cordless drill.

For strength, and to make sure no cut ends poke out beyond the edges of the work, I cut the 6mm rods about 100mm long so I can bend them around the (normally thicker) reinforcing of the edge, using the grooved pipe benders (see below).

Cutting reinforcing to length

To mark reinforcing rod: use engineer's chalk or nick with grinder, or hacksaw. Or tape with insulation tape. Or have a pre-marked length of wood to cut to.

To make the cut, use bolt cutters or angle grinder (with cut-off disk which is thinner and so faster) for 6mm and under, and angle grinder for 10mm and over. There is a big bolt cutter which can cut thick reinforcing, but it costs hundreds, and takes a lot of effort to use. The angle grinder only costs a hundred or so and is quick and versatile.

When cutting with angle grinder, make sure the rod doesn't sag downwards as it weakens and jam the grinding wheel. Watch the cut ends—they will be sharp-edged as well as almost red-hot at first.

If you have a drop saw, you can put a metal cut-off wheel on it and use it to cut reinforcing quicker still, but only if it is straight and can be brought to the saw.

Bending reinforcing

Big curves

Unless you have a pipe bender, as engineering workshops often have, gentle even curves are best made by fixing one end by clamping or stake in the ground etc, then walking the rod around a curved object like a water tank, culvert, or a series of stakes in a curved line. Failing that, you can work the rod through a two-rod bender (see tool list), bending a short length at a time.

And once you have the first shaped rods in place, you can bend the thinner ones over these and the natural spring of the thin rods will make a surprisingly smooth curve.

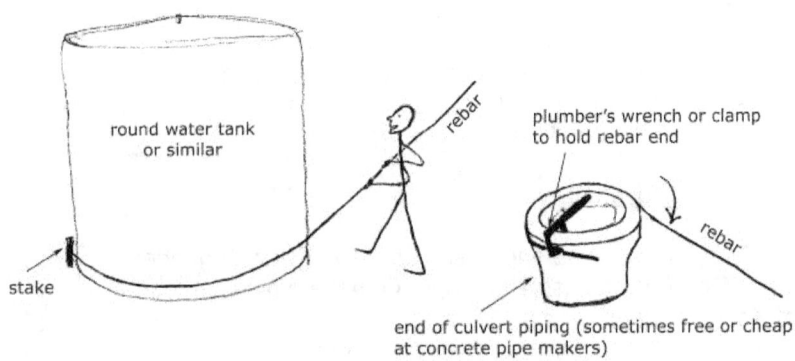

FIGURE 18 Ways to make smooth big curves

Tight curves

These can be made with the two-rod bender, or, for really tight bends, there are lever benders you can buy. The ones that do thick reinforcing are expensive, though. They can be hired, if you have a lot to do. I usually just work with the gentler bends I get from my two-rod bender tool.

Another way to get a tight bend is (if it is possible) put the reinforcing into two pieces of strong steel pipe, 20-30mm diameter. Put the ends of the two pipes together at the place you want to bend, then stand on one pipe and lever the other up.

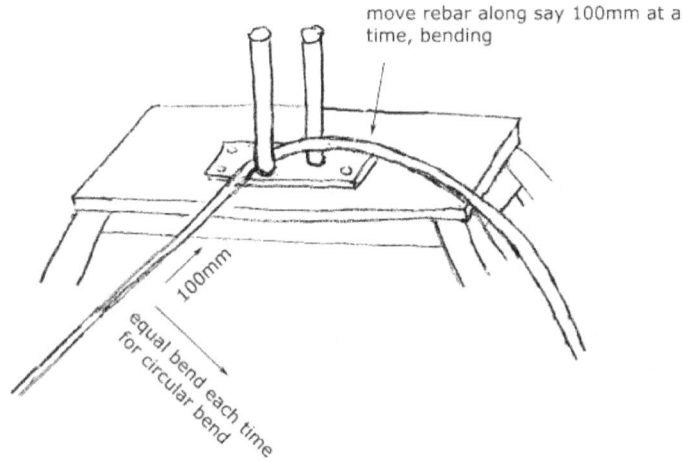

move rebar along say 100mm at a time, bending

100mm

equal bend each time for circular bend

FIGURE 19 Curving by hand on two-bar bender
Bending thin reinforcing around thicker, and re-bending bar already in place

One of the handiest tools of all is the grooved pipe. This can bend 6mm rod sharply or gently, and even wind it tightly around a thicker rod. For adjustments to thicker rod, I use the three-pronged bender.

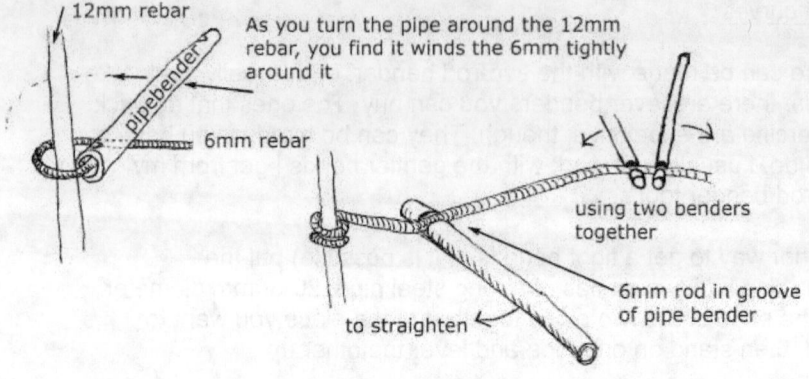

FIGURE 20 Bending reinforcing bar, or rebar

<u>Tying reinforcing</u>

Bag ties are quick and handy, but have a bulky 'tail' after twisting, that needs to be laid flat, and they can allow rods to slide. Can use two baggies diagonally to each other to minimize this. Short lengths of thick lacing wire are also good, though slower. Thin lacing wire is fine for thinner rod. Wind wire diagonally both ways like rope lashing to really tighten to avoid slipping of crucial joins (before plastering— after it doesn't matter as the plaster holds everything in place).

Welding is a whole other ballgame, and is not recommended by some as they say it can cause corrosion and weakening of reinforcing. But for crucial complex joins, like where the ribs of a dome join, it can be very handy. Tying, though, is the preferred method for ferrocement. Note too that any flexing is removed when the chicken wire is laced on, and then when the plaster hardens, everything is embedded in the rigid plaster, including the tie wires.

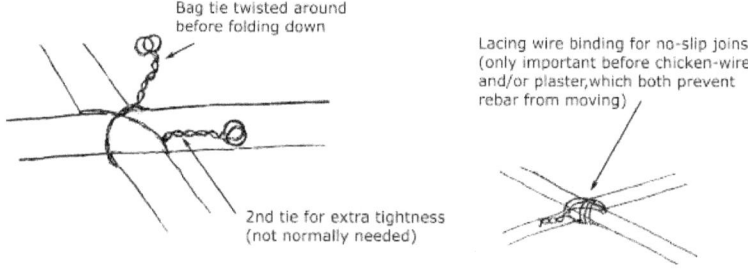

Bag tie twisted around
before folding down

2nd tie for extra tightness
(not normally needed)

Lacing wire binding for no-slip joins
(only important before chicken-wire
and/or plaster,which both prevent
rebar from moving)

FIGURE 21 Tying rebar

Zig-zagging to make ribs/beams

Thick lacing wire of 6mm rod can both be used. Simply tie one end
to one of two parallel thick rods, angle between the rods, tie to other
parallel rod, bend to angle back to the first rod, and so on,
zigzagging all along the gap between the rods, making a rib or
beam that can strengthen an opening or a wide span e.g. on a roof.

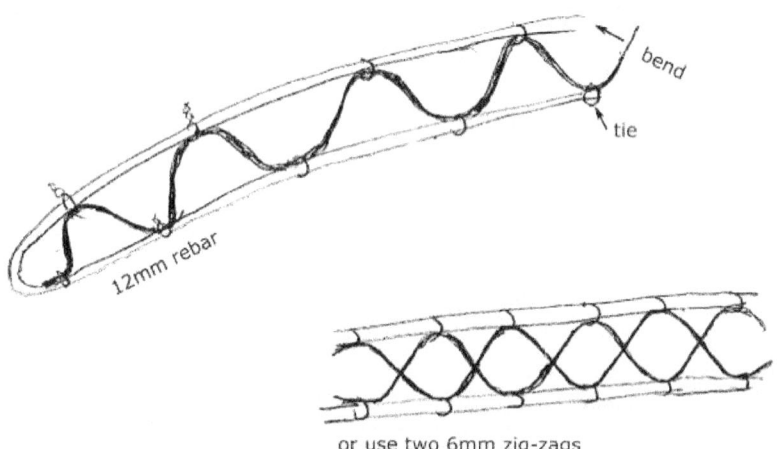

bend

tie

12mm rebar

or use two 6mm zig-zags

FIGURE 22 Zig-zagging

39

Lifting-loops, conduit, holes

Lifting-loops

These are always put around the bottom of concrete tanks for lifting and moving. It is a good idea to put some in your work if it is too big to lift without a crane. Use 10 or 12mm reinforcing rod, overlapping plenty with other reinforcing.

Conduit

This is for any fixed wiring your work may need. If in doubt, put some in, before the chicken wire. It is a lot easier than doing it later! Ask a plumber's supplier for details—you can cut and bend pvc pipe or buy ready-made elbows. There is a special glue that joins it.

Holes

To make small openings for e.g. crystals to go into, force the chicken wire apart by chisel, hammer etc, and plug the hole with a short section of plastic piping or similar before plastering. Bend the chicken wire back and if necessary lace it down. When the plaster sets, knock the plug out. Or, if it is plastic pipe, you can leave it in.

D: Chicken Wire

This is the stage that is most routine, in a way, but takes the most actual time, though it is mostly light work. But be warned: it does take patience. Each square metre takes about 10 metres of lacing, or maybe 1000 stitches. If you can get a buddy it is a great stage for two or more to help, one on each side of the mesh. You can talk, pace each other, and the time doesn't drag as it can when lacing by yourself.

Cutting chicken wire

I strongly recommend using an angle grinder (with the thinner steel cut-off wheel preferably, though a grinding wheel does work), for zippy speed and less pain to your hands—using snips your hands are moving close to the cut mesh and always get scratched and poked, unless you wear thick leather gloves which are clumsy and tiring. There is quite a lot of cutting of mesh in ferrocement work.

When measuring up, allow plenty for the curves, though especially sideways there is some stretch in the mesh.

Especially for longer pieces, find a level place where you can unroll the chicken wire. Put a weight on the end to hold it down, then lift the mesh a few inches where you want to cut, and sweep across it with the angle grinder in one hand, holding the cut end with the other. You can lay out and cut four or more thicknesses at once, too; for example when you need to make wedge-shaped pieces for a domed shape. Just make sure there is something like sand or gravel or layers of wood or cardboard underneath, as the cut-off wheel will of course cut through the bottom layer of mesh.

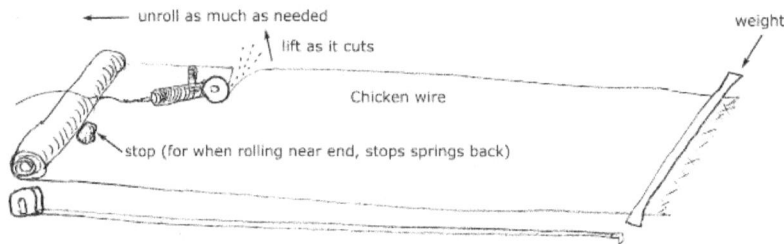

FIGURE 23 Cutting lengths of chicken wire

Lacing chicken wire in place

This is the single most time-consuming stage, as I have said. But it is satisfying, and not hard if you are patient and methodical, and use a few simple tools and techniques. It helps to have a helper, too, for the other side of the framework. In boat building, they used short wire ties, and twisted them tight then bent them under the mesh so they wouldn't stick up and spoil the plastering. We tried this, but it is very time-consuming. I developed a 'sewing' method, which is much easier and more fun.

Make a bundle of 2-metre lengths of lacing-wire, using a drum winder (see drawing). Put the bundle over a wire hook and hang it up in a handy place on the framework.

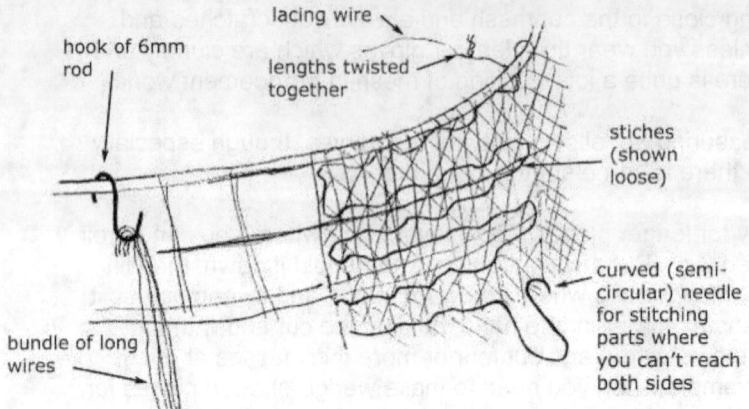

hook of 6mm rod

lacing wire

lengths twisted together

stiches (shown loose)

curved (semi-circular) needle for stitching parts where you can't reach both sides

bundle of long wires

FIGURE 24 Lacing chicken wire in place

Then place your first cut lengths of chicken wire on either side of the framework (usually two layers per side. Line them up roughly (the meshes don't want to be perfectly aligned anyway—the opposite is best for reinforcing and holding the wet plaster), and put an end of lacing-wire through. Now, depending whether you can reach both sides, you either use the semicircular needle from one side, or just thread the wire in and out of the chicken wire layers by hand. If there is a curve, especially a concave/convex one as in a pot or dome, the stitches will need to be close, about 50 mm apart, but in flat areas you can get away with about 100mm. Space the rows of stitches the same—50-100mm apart. Often you will end up crossing the lines. It doesn't have to be neat—it all gets covered!

It is quite like hand-sewing. Pull to tighten every one or two stitches. When you get to the end of one piece of lacing-wire, twist another onto it and continue. When you have fastened one area of mesh, tie off by making a few very short stitches and then snip the lacing-wire and bend it into the mesh.

Each area of mesh needs to overlap the others by say 50 to 100mm, and to be stitched through the overlap so there are no areas that are loose—this is vital, to make sure all the mesh is pulled down well onto the reinforcing, so there are no 'baggy' bits. There is a way of final tightening, using needle nosed pliers or similar, and twisting/deforming the chicken wire in loose areas, which can tighten it up quickly. But this method doesn't work on concave areas—only stitching can pull the chicken wire in there.

42

Pay special attention to cut edges of the mesh—these often tend to stick out so need extra stitching.

If you have an odd area to cover, use offcuts of chicken wire, as long as they overlap. It is inevitable that it will feel like a patchwork, if the shape is at all interesting! It is surprising how as you stitch it all starts to pull into shape and look good. And the whole framework will start to feel very firm and strong, even before plastering.

E Plastering

Mixing Plaster

Above all, don't put off the plastering through anxiety and fear of failure. Just begin (after preparing well and allowing plenty of daylight time including about half an hour for the clean-up) and you'll find it will work out fine, and you'll actually have fun! This is almost the quickest stage, and really dramatic, and you don't have to be a perfect plasterer to do it well! I have had 'real' plasterers help us, and they do it just like us, mixing the ingredients and plastering the mix on the wire.

Once the framework is all wired and all loose ends of chicken wire (or reinforcing) are bent down, and there are no 'baggy' areas, and all the reinforcing has chicken wire over it, it is time to get the area ready for plastering. If the work can be moved to a convenient place for plastering and curing, and won't be too heavy to move back when it is plastered, move it now. It is too late when the plaster goes on. Consider where the runoff from the plaster mixing and clean-up will go, and avoid blocking drains or filling gardens with cementy silt! Lay a cheap tarpaulin or drop sheet under the work, if there is soil or grass under it. Then you can scoop up and reuse all the bits of plaster that drop off—there is a lot of this sometimes, especially when you are practising. Make sure you have a good supply of clean water, from a hose preferably, with an adjustable trigger nozzle on the end.

I set up the bags of cement on wood to keep them dry, and high enough so the lowest bag can still pour out into my measuring buckets.

Have a knife or similar to cut bags open - see drawing. Since most people will have access to a concrete mixer in NZ, even if they have to hire it, I will describe machine mixing mainly—hand is just a matter of scaling down the batch sizes so it's not too hard to mix it all thoroughly, and then just hard work with a shovel.

There are different schools of thought on whether to mix dry first, but I have found it quickest to put the bulk of the water in first (not all of it, or you might get too much—leave some to fine-tune the consistency). Here is a typical strong (2:1) mix for a first coat— sometimes second coats have polypropylene fibre and can have less cement to reduce cracking.

For an average concrete mixer:

- Around half a bucket of water to start (more or less depending on how dry or wet the sand is), then add plasticiser (two plastic milk bottle capfuls, or 20 to 30 mls) and oxide colouring if you are going that way (but see the section on Painting for comments on this).

- Then a 10 litre bucket of cement thrown in

- Then three 10-litre buckets of sand.

- Then another half bucket of cement. I do this partly so the last buckets of sand get some of the cement quicker.

- After a minute or two, when you see how wet or dry the mix is going to be, gradually add more water with the hose, using this extra water to wash any cement around the edge of the bowl into the mix. If it started to 'ball up' early on, this is because there was not enough water to begin with; just promptly spray water on to break up the balling. If once it balls up tightly, it takes a long time and extra water to smooth the mix out again.

- If you go too far and the mix looks not like smooth thick porridge but thin runny porridge, don't worry; just add some of the driest sand you have, and some cement as well to whatever ratio you are using.

Applying Plaster

First Coat

Make sure you have a lovely light, well-mixed batch of plaster, of the right consistency so it goes through the layers of mesh like toothpaste, when pushed or trowelled on (you could take out a

44

handful of the mix and try this test).

You are now ready to plaster. Start in anywhere that makes sense (don't paint yourself into a corner!), probably at the bottom, so that any falling bits of plaster don't get lodged loosely in the mesh and begin to dry before you get to that part. If you go bottom to top you can always smooth over any parts that get dropped on, later.

Work the plaster into the mesh well, in several quick light passes rather than one heavy one. Push plenty of plaster over the mesh each time so it goes right through and oozes out the other side. Don't
'worry' the plaster once it is on (in); the more you move it around, the more the water floats out, carrying the cement with it, and you get running plaster as well as sandy dry plaster.

Aim to leave just enough plaster on the surface of the mesh to cover it all—say 2 to 5 mm cover (it will vary with the undulations in the mesh). If you come across wire or mesh sticking out, use the edge of the trowel or similar to gentle hit or push it back under the plaster. Or if you find a 'bag' in the mesh, use needle nosed pliers to twist the mesh tight .

Until you are experienced, it is best to assume there will be another coat, even if you don't plan to carve it; so on the first coat just focus on pushing the plaster into the mesh with no air gaps, then roughly but evenly smoothing it over (not smooth-trowelling it; remember you don't want a smooth, cementy finish if you are doing a second coat).

Smooth the second side, putting on a little extra plaster if needed, especially if the plaster from the first side hasn't oozed right through. On surfaces like ceilings, it pays to wait until the plaster has begun to go off before smoothing the second (under-) side. This stops the plaster from dropping off the underside. Another trick is to add a handful of fibre to the batches you use for the undersides—it helps the plaster to stay up. And the rule about not overworking applies to undersides especially; if you do it the plaster will literally drip off.

If you plan a second coat, as soon as the plaster 'goes off', i.e. starts to be a bit brittle, and not spreadable without crumbling, scratch the plastered surface with the Scratcher, to leave grooves about 2 mm deep. A Stiff broom could be used, but with the curves of ferrocement it isn't the best—it would be easy to gouge into the

45

soft plaster with the broom, or miss parts entirely. Another approach which I have used is to leave the surface as is, and before the second coat paint the entire surface with concrete adhesive (Aquadhere, Acrylabond). This is not as strong as a mechanical 'key', and it is expensive for big areas, but definitely better than nothing. Your second coat will delaminate if the undercoat is smooth and you don't use an adhesive.

Thickness of the plaster is everything in plastering, and if it seems too thick and is hard to push into the mesh, add a little water and work it into the batch in the wheelbarrow. It is amazing what a difference just a little water makes to ease of working.

Second/final coat

This can be 3:1 or even 4:1 instead of 2:1, and should be a little runnier, so it can go on smoothly and thinly. The dry plaster underneath seems to always absorb some moisture from the second coat, too, although not if it has had a thorough coat of adhesive, if it is newly wetted down. Always spray down before doing the second coat, but don't leave any pooling or really wet surfaces, or the plaster will slide off it! I find it easier to leave the first coat almost dry, and err on the side of wetter plaster which will then have some of its moisture sucked out of it. This sucking actually helps to make the new plaster make good contact.

Third/carving coat

This can be thick, and you can build it up even thicker in places for e.g. pillars, carved vines, flowers, lips on edges, etc. I always use fibre in the mix for this coat, and often add some adhesive in the actual mix; and make a drier mix. I always put the fibre in first after the water, so it disperses and no clumps are left.

This produces plaster that can be held in a big ball (say 200 mm diameter) in one hand. It can be stuck onto a wall, to form a half-pillar (usually it takes two coats to form a pillar that sticks out say 100mm). It can be smoothed and pressed like clay, then of course carved when set. I tend to still go 2:1 in the mix, so it is not so crucial to cure it for seven days wet curing; even at less than optimum strength, with that rich a mix it will still be any amount strong enough.

46

Build up plenty to allow for carving the highest points. Carving can be started within a few hours depending on the temperature, though if the plaster is too soft the fibre tends to pull out.

Sponge finishing

If you want a smooth but gritty, sandstony finish, sponging is a good way to go. Wait for an hour or so, or until the final coat is going off, then with a wet stiff sponge (or rag) lightly rub in a circular motion over the whole surface. The plaster should be just fresh enough for the sponge to move it a little and even out any trowelling marks, at the same time exposing the sand in the plaster and getting rid of the shininess of the cement 'skin' that always forms when plastering. Every now and then you will need to hose and wring the sponge, to remove the build-up of plaster in it.

When plaster has been sponged it seems to dry out more quickly, so will need even more care to keep it shaded and not let it dry completely before it is fully cured.

F: Curing Plaster

This is essential for strength, for the first plaster coat especially, and takes discipline. Ideally, cover the new plaster from sunlight and keep a constant wetness over the whole area, taking special care not to let edges dry out. If at any time, the plaster goes whitish-grey, before it has been curing at least three days, preferably seven, this is very bad—it means it has dried out and it will never cure properly, even if wet again straight away. And too-quick drying also causes cracks in the plaster.

There are two ways around wetting the plaster continuously for a week, and that is to completely coat the new plaster with a sealer (e.g. with concrete adhesive), or use a lot of adhesive in the mix so it doesn't dry out nearly as quickly. But this is expensive and sealing is time-consuming. It is handy though, if you have only plastered a small area (repairs, additions etc), especially if you don't want water dripping over it for a week.

The carving coat is not so crucial to keep wet for so long, as it doesn't have to be as strong, not being the structural part, the ferrocement layer. But it will be the part subject to knocks and abrasion, so it is still better if it is hard.

I use a hose with holes I have punched at 100mm intervals, on all sides of the hose, using a sharpened 40mm nail in a square of wood. The hose is blocked off at the end. Then I drape it carefully over the finished work and turn the hose on gently so it just trickles from the holes. Ideally I try to also cover the work with a blanket of hessian to keep the sun off. Then the hose can go over the hessian and keep it wet.

Of course, for small enough objects, you can simply put them in a tank of water or a pond etc, for a week. Bear in mind that the alkalis coming off curing concrete can kill fish (if there's a lot of concrete and not much water. So, for example you can't put fish in a ferrocement pond for the first week or two after plastering it).

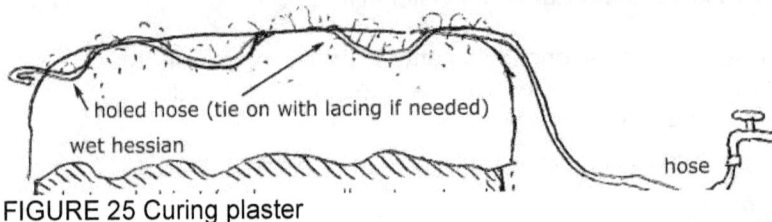

holed hose (tie on with lacing if needed)

wet hessian

hose

FIGURE 25 Curing plaster

G: Scraping , Sanding and Carving

Using the tungsten scraper, I have found I can carve quickly and in fine or coarse detail, especially if the plaster is less than 24 hrs old. After that it gets a lot more physical to scrape the hardening plaster. But the tungsten can carve even completely cured plaster, and do hours of work before needing to be sharpened or replaced. I carve dry usually, but you can carve wet, washing away the scrapings. The main action is a pulling stroke, flat for big areas, angling to use the corner of the blade for grooves and lines. Be careful not to carve right down to the mesh, which will then rust (though it can be scraped down and painted over).

The grinder is good as a last resort, but you need to deal safely with the fine concrete dust.

For final smoothing and rounding, I use coarse sandpaper, wet and dry usually, though if you don't wet the work you can use ordinary sandpaper, and slap the folded paper on something like an old chair back to unclog the grit. I use 60 grit, 80, 100, and even finer for delicate carvings.

If you are carving plaster with polypropylene fibre in it, you can sand off the fibre or use a gas flame to singe it off, before painting.

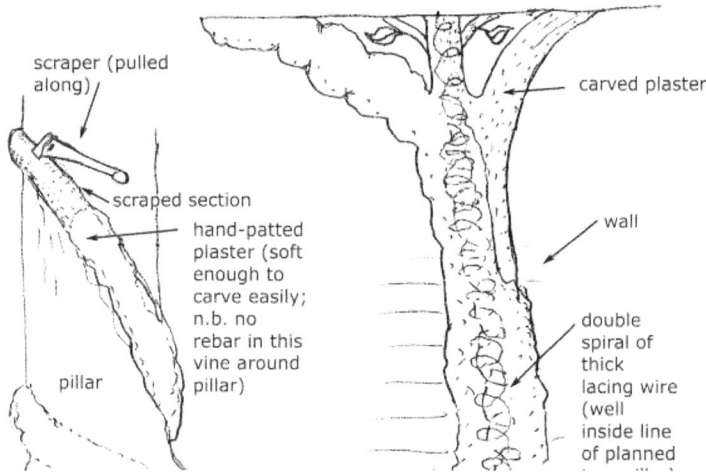

FIGURE 26

PAINTING

Paints and finishes are a huge topic, and there is so much technical development all the time that it is impossible to give a complete rundown of what is available or what to use for what application. But in general:

The painting is best left a few weeks until the plaster is cured and all the 'salts' have come out of it. Then a good acrylic will do well. Sand between coats to get rough bits like bits of fibre or grit.

I use a self-priming white as it is a good base for vivid colours. If you cant wait, there are special but somewhat dearer and harder-to- find paints that can be put on over 'green' plaster that is only a few days old.

You can do a wash of thinned-down acrylic over the base coat, just as in decorative finishes on wood or wallpaper. Then coat over the top with a good varnish. There are water based varnishes that dry fast and don't smell. You don't need a special concrete varnish if you are coating over the acrylic.

The plaster can also be stained (ask at paint shops, but I use acrylic watered down about 5 of water to 1 of paint), then coated with a concrete varnish.

Or, of course, you can use oxide colouring additives in the plaster mix. But these are expensive and never give vivid colour when used with the normal grey cement, and it is hard to get total consistency between batches of plaster. You can buy white cement for vivid staining, but it is much dearer. The other point is that for durability in the weather, ferrocement should be painted, as no plaster, even with fancy additives, is completely impervious to water, and tiny crack can let in more. So a good acrylic roof paint or would be best for ferrocement roofs; or even a special 'membrane' paint that forms a tough thick membrane over the substrate, and is waterproof even if there are small cracks beneath.

I Cutting and Drilling Ferrocement

Drilling

Often you need to drill the finished work, to put bolts through for various reasons, or to put plugs in to take screws. This is usually easy, using a tungsten-tipped drill bit. Sometimes you hit the thick reinforcing; then it is best to re-drill nearby. You don't want to weaken the reinforcing, and it will damage the tungsten tip to try and drill through bigger steel. (tungsten does handle thinner reinforcing and chicken wire all right though). Use galvanised bolts and grout the hole with a rich cement-sand mix, preferably with concrete adhesive in it too.

For screws, use the plastic plugs which you push into a hole sized for a tight fit when the screw goes in and expands the plug.

Cutting

Planning should avoid the need for this, but things come up.... The simplest solution here is the angle grinder. It will wear out a few cut-off disks, and make a lot of dust, but it is quite quick. If you have a good jigsaw, and buy a tungsten blade for it, you could cut it with this, but if you hit larger reinforcing you would have to grind this through, or switch to a metal cutting blade. But the plaster would blunten this quickly. I have never done this (yet), but for curved cuts it could be worthwhile trying.

You can hire big drills with circle cutters which cut out a plug about 50mm diameter, but if you hit big reinforcing, the tungsten teeth can be broken (I know from experience!).

Another approach is to chisel away on both sides, possibly with a power chisel, until the reinforcing is exposed, and then grind or jigsaw that.

After cutting, of course you will need to plaster, or at least paint, over the cut to protect against rust.

J DOORS AND WINDOWS , STAINED GLASS , & GLASS PEBBLE WINDOWS

This is a big topic, and I am not an expert (yet). But we have done a few doors and windows in Dreamspace.

Windows

Small windows

I have made non-opening porthole windows by plastering thickly around a round opening then pressing the 6mm circular glass into the soft plaster (if the glass breaks, the plaster will have to be scraped off). Or, scrape a rebate for the glass in the still-soft plaster, and bed the glass in with a bead of silicone suitable for concrete/glass.

Plan ahead, and make oversized openings in the framework where you want the windows, allowing for the usual plastering (20mm or so all round, i.e., 40 mm added to the width, 40 mm to the height).

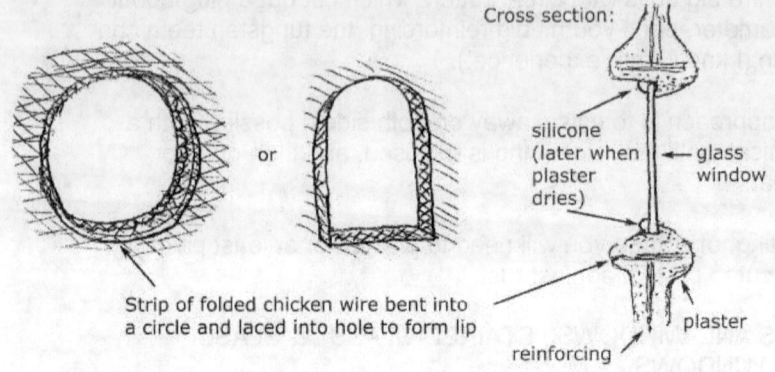

Cross section:

silicone
(later when
plaster
dries) ← glass
window

plaster

reinforcing

Strip of folded chicken wire bent into
a circle and laced into hole to form lip

or

FIGURE 27 Small windows

Tiny rainbow windows

For tiny windows that let in light and make rainbows on the wall, I
have cut circular holes in the ferrocement and silicone facetted
crystal into the holes (see under Cutting ferrocement). But it is best
to plan ahead where you want the holes, and just force the chicken
wire apart by chisel, hammer etc, and plug the hole with a short
section of plastic piping or similar before plastering. Then when the
plaster sets, knock the plug out.

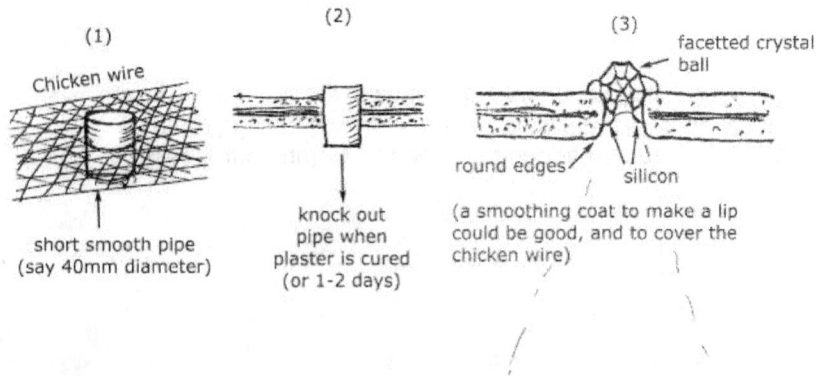

(1)

Chicken wire

short smooth pipe
(say 40mm diameter)

(2)

knock out
pipe when
plaster is cured
(or 1-2 days)

(3)

facetted crystal
ball

round edges

silicon

(a smoothing coat to make a lip
could be good, and to cover the
chicken wire)

FIGURE 28 Tiny windows

<u>Larger, opening windows:</u>

Here, again, the best approach seems to be to make an oversized opening, with a ferrocement strip around it to make a width of say 150mm, then plaster the window frame in place later (the frame is what the window sash, the opening part, fits in and is hinged from). Allow the usual 20mm all round, plus about 15mm (so total 35mm on all sides i.e. 70mm extra width, and 70mm extra height) to allow room to push in a good layer of plaster (with plenty of fibre in it) around the frame.

It is best to wind a couple of layers of waxed plumber's tape around the frame. This tape is for lagging of any pipes that go through concrete, to seal around them and allow for expansion/contraction. I have done temporary windows without this tape and gaps have appeared.

When fitting the frames, position them with pebbles or chips of plaster, or screws through the frame. I use self-tapping wood screws—quick and easy!

I made my own frames out of macrocarpa, and the outside sloping sills I made of ferrocement. I have also cast fibrous cement frames, with a semicircular non-opening top, and fibrous cement windows to fit, but this was quite a mission and isn't worth it unless you are making a lot of windows.

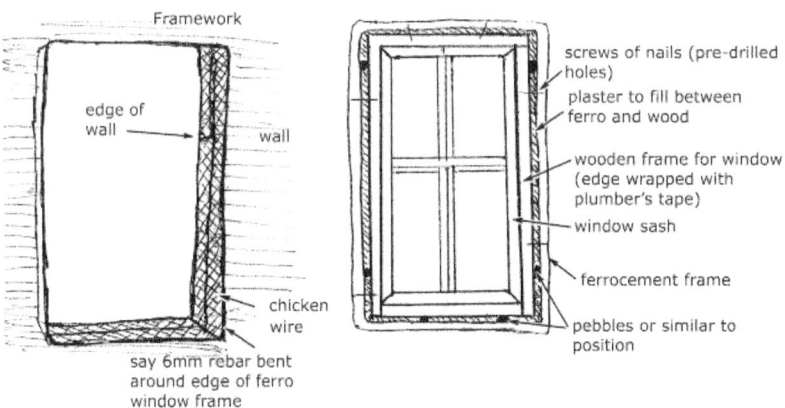

FIGURE 29 Windows in ferrocement

53

Doors

These look great in domes if they are round at the top. We used slabs of macrocarpa, jig-sawed to shape. For hinges, we have got special ornate ones cut out of plate steel for one door, but for another we have used wooden frame and ordinary hinges. Two we made with farm gate hinges—it looks rugged and good, but it's not ideal for precision fitting.

So the simplest thing is to make a rectangular oversized opening in the ferrocement, as for the windows, and plaster a wooden frame in. I would put a few galvanised bolts through to hold the frame in place securely.

For a round-topped door, the wooden frame can only do the sides, so if you need a rebate at the top, it can be done after the door is fitted, by plastering over the inside wall to build up the rebate. Use a piece of cardboard or the fluted plastic they make real estate signs out of, as a mould around the curve of the door.

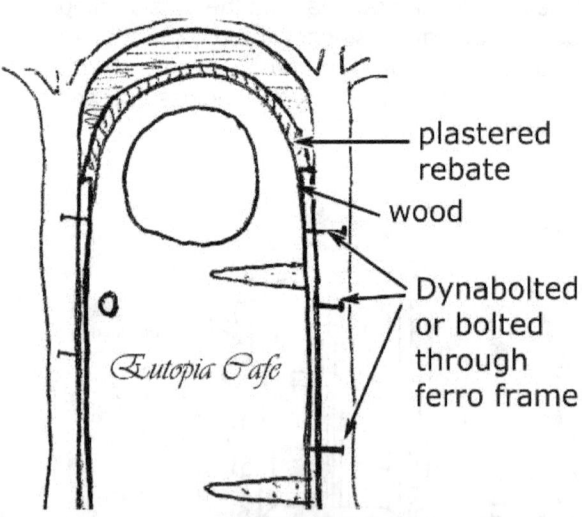

FIGURE 30 Arched door in ferrocement

2012 addition: Ornate plywood doors as seen in the courtyard photo below. If you are doing just one or two, jigsaw out the pattern you have drawn on the ply, and rout around the window sections with a rebating bit with a ball bearing end. If you want to do more, it

54

is worth using your first one as a routing pattern:

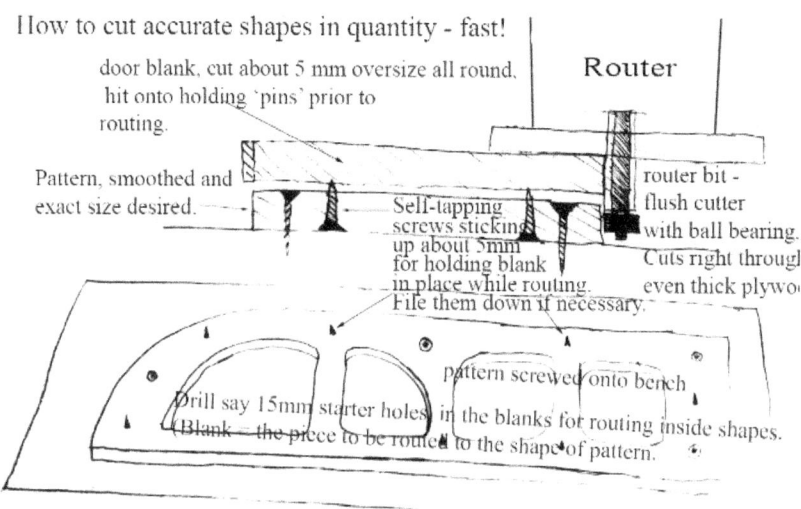

How to cut accurate shapes in quantity - fast!

door blank, cut about 5 mm oversize all round, hit onto holding 'pins' prior to routing.

Router

Pattern, smoothed and exact size desired.

Self-tapping screws sticking up about 5mm for holding blank in place while routing. File them down if necessary.

router bit - flush cutter with ball bearing. Cuts right throug[h] even thick plywo[od]

pattern screwed onto bench

Drill say 15mm starter holes in the blanks for routing inside shapes.

Blank - the piece to be routed to the shape of pattern.

This is the method I used to make thousands of photoframes, in the olden days (last Century) when I was a manufacturer! So don't let the hardware man tell you that you can't cut right though ply or wood with these (tungsten carbide tipped) cutters with ball bearings - you can. You just guide the router firmly along the pattern (anticlockwise on the outside and clockwise inside the window holes), set so that the ball bearing is running along the middle of the pattern, but not so low it digs into the bench below!

Stained Glass and Glass Pebble

I have done some traditional stained glass, and although it is satisfying, it is expensive, difficult to do well, time-consuming, and uses lead which is a cumulative poison. So I looked at different alternatives. The simplest I have found is to cut the pieces of glass and thinly clear-silicone them onto a plate glass window, and then use black grout between the pieces.

Or, my favourite is to use coloured glass 'pebbles', or even round marbles, siliconing them on – just with a bit more silicone so they don't roll off! The only limitation is the colours available—not as many as for sheets of coloured glass. I have used both glass

55

pebbles and coloured glass on the same window – the height difference doesn't matter.

Glass 'pebbles' glued on with silicone.

Black acrylic roof paint mixed with plaster of Paris to make a putty, wiped over the glass, excess wiped off.

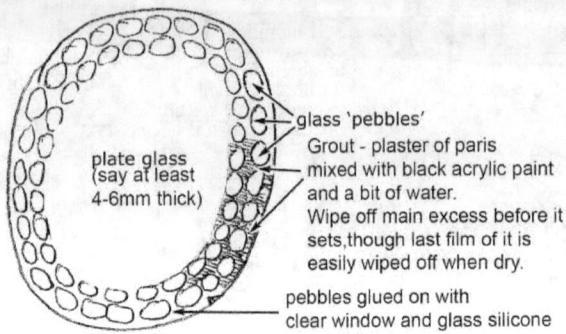

glass 'pebbles'
Grout - plaster of paris mixed with black acrylic paint and a bit of water. Wipe off main excess before it sets,though last film of it is easily wiped off when dry.

plate glass (say at least 4-6mm thick)

pebbles glued on with clear window and glass silicone

FIGURE 31 Stained glass with glass pebbles
A great grout

After trying normal black tile grout, and black putty, I found a much quicker and cheaper method, which is to mix some ordinary plaster of Paris (gypsum plaster/casting plaster) with enough black acrylic roof paint (thinned with a bit of water) to make it into a 'putty'. It dries slower than Plaster of Paris mixed with water, so there's time to work it into the gaps of the glass and wipe it off. I usually use rubber-gloved hands for the bulk of the wiping off, then paper towels when it has started to set. Even when set you can rub the last of the mixture off the surface of the glass, with a damp cloth or paper towel.

The result, especially from the inside, is lovely. The pebbles look like glowing jewels. I have even tried it outside on a bay window fully exposed to the sun and rain, and it is standing up well years later. The finish is semi-gloss because of the roof paint.

If you want the same texture inside and out, try sticking matching pebbles on the other side too – this will double the darkness of the colouring of course, but often the glass pebbles are not very darkly coloured anyway.

As with all the techniques in this book, I don't claim the last word on it, and would love to hear from you if you know of a better or

56

additional way, to push the limits of what we can do in ferrocement, glass and wood.

Stained glass lampshades

If you get a lamp with a glass or plastic globe, you can use this technique to make a beautiful 'Tiffany' style lamp – VERY easy and quick to do! For the best effect use clear, not frosted, glass or plastic, as the frosting dulls the effect.

Decorative rocks bottles etc.

You can easily push apart the chicken wire to place a rock or bottle etc into it, then plaster it in place.

Polystyrene shapes

A slightly messy but effective way to make a smallish cavity in concrete, is to plaster over a lump of expanded polystyrene (the very light white stuff goods are often packaged in). Then when the concrete sets, dig out the polystyrene. We have used it for cavities inside solid pillars:

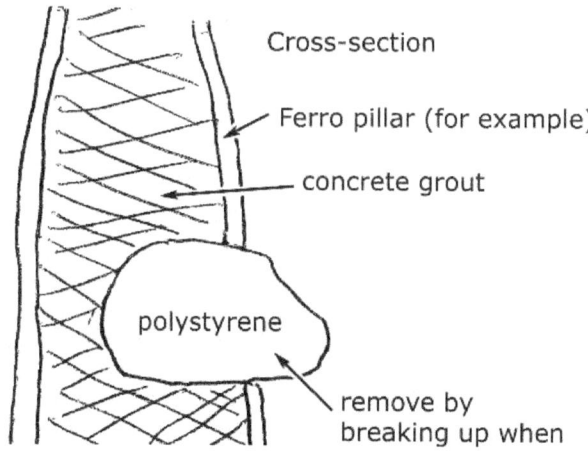

FIGURE 32 Polystyrene cavity

Or, use it as a form to plaster over to make e.g. pillars. The polystyrene won't rot or crack the plaster, so it is ideal for this.

Shelves and benches/sinkbenches/desks

It makes sense to use ferrocement (OR plaster – see next section) for these, especially if they can be permanent fixtures. The curves normal in ferrocement make wooden or steel shelves and benches harder to make.

So, if you are building them in from the start: Do the walls first, complete with chickenwire, then make some L-shapes in 6 or 10mm rod, making the L as long as the shelves are wide. Slip one arm of the L's vertically into the chickenwire. Run two lines of 6mm rod along the outline of the shelf (or use 10 to 12mm for a bench), and more along the middle for wider shelves or benches, and some crossways. For extra strength, wind the ends of some of the rods around the edge rods, or make L's in the ends and lace them on. Drop legs of one or two 12mm rods wrapped in chickenwire, with an L at the top for lacing it onto the bench. Legs need to be spaced about as they would for a wooden table – say 1.2metres apart. For a bench you want it to be perfectly level, so go to a little extra trouble to reinforce with straight rods, or welded 665 mesh, so you don't have to load the top with too much plaster to get it even – the more plaster on top of the chickenwire, the more likely it is to develop cracks – there is a lot of stress on a tabletop. Use fibre in the second coat, and a level to get it level or very slightly sloping if you are putting a sink in (this is not hard – use stainless and drill holes around lip of bench and lace it in place)

Wait for the second coat to get the edges really smooth and even. Use plenty of fibre, and hand-smooth with latex gloves as usual. Final smoothing is with sandpaper, and scraper if needed, when the plaster has set – after a few hours or overnight.

Polystyrene-core shelves/benches

If you are adding the shelves or benches later, there is another way to go: you can use polystyrene sheets, say 20 or 30mm thick, cut to shape and glued together (using waterbased construction glue from a cartridge) to make shelves and benches, then plaster them, and lay a plastic-coated fibreglass mesh into the plaster. This is how they make the cladding-plus-insulation in some houses, and is quite strong and light. For fussy shapes, and ones where you want a level surface, as in shelves and benches, polystyrene has a lot going for it. Also, if you have to take it away, it will be a lot easier to demolish than ferrocement. It helps to have a hot-blade polystyrene cutter, like plasterers use, but a sharp knife or saw will do. Get

some plastic-coated fibreglass tape, and enough metres of fibreglass cloth reinforcing to do the area
- one layer should be plenty. The tape is for attaching the shelves to the wall – see drawing. Use a cartridge gun with water-based construction glue to fix the tape to the wall and the polystyrene.

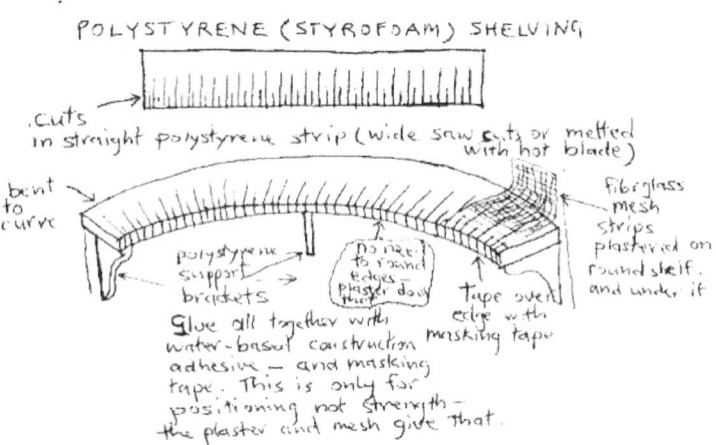

FIGURE 33 Plastered Polystyrene shelves

By making cuts most of the way across the poly shelving, say every 25 mm, you can bend the shelves to fit a curve. Then I taped the curved edges with masking tape. The mesh and fibrous plaster gives the strength, so the cuts don't matter. Also, for interior, I've found casting plaster (gypsum plaster – plaster of Paris) very quick and easy. For extra water-resistance and to slow the setting, I used the usual Cemix Acrylbond in the mix – about two parts water to one of acrylbond.

If you want a really thick shelf, use 40 or even 60mm polystyrene.

You can also use it as a form to plaster over to make e.g. pillars. The polystyrene won't rot or crack the plaster, so it is ideal for this. (Otherwise, use the spiral wire method - see FIG. 44 below.)

K Moving Ferrocement Objects

There is a tendency for the ferrocement project to get heavier, when you add coats and carve it. The extra coats can be more fragile, and chip off if the whole weight of the tilted object rests on one bit. So I have found it is worthwhile finding a piece of ply, or old carpet, a flattened carton, etc, to 'walk' the finished object onto and then drag. If it is too heavy for this, forklifts are good and often surprisingly close by, in factories and warehouses. Talk nicely to the operators, and perhaps for a few beers they will shift your work for you.

Also hiab trucks have cranes that reach sometimes many metres and can lift tons. They can be hired, current rates maybe NZ$100 per hr. If there is no convenient place to put lifting chains around, you can drill two holes say 150mm apart through the ferrocement and put a loop of 10mm reinforcing through the holes, and hammer them over on the inside, overlapping the ends so you can lace them together. Then bend the handle up at an angle, and it will take a lifting hook. You will probably need several of these for lifting a big object safely.

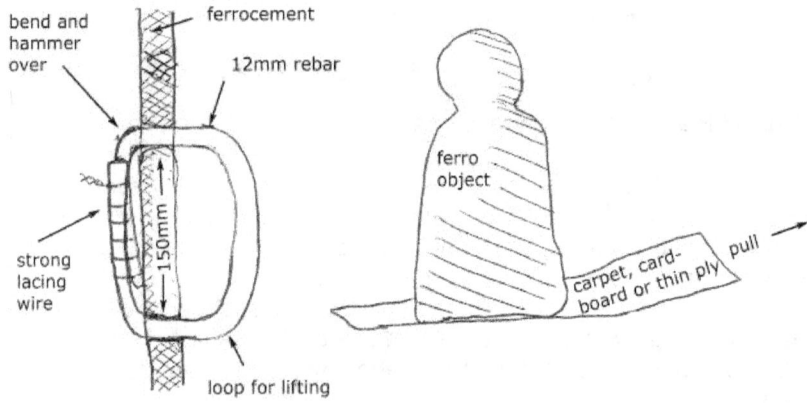

FIGURE 33 Lifting loop and sliding method

Ideally, you will have built these loops into the object from the framework stage.

For getting a heavy object into a trailer or van, try two thick beams of say 50 by 150mm wood, spaced like rails, long enough, say 2.4metres, to make a ramp up which you can walk and slide the object. Safety first, though!

60

6 Sample Projects

Project 1. Garden border for raised beds

The simplest way to make these is to get a manageable length of chicken wire and fold it lengthwise twice, to make a 225mm wide strip of four thicknesses, then curve it around the area you want, adding extra lengths as needed, overlapping and lacing together. Or fold lengthwise into three, to make a strip 300mm wide, still strong enough.

For permanence it is best to make a little channel, say 50mm wide and 30mm deep, in the ground, and put a line of plaster in it, and lay the chicken wire strip in it, so the bare wire doesn't touch the ground and eventually rust. Try and make the line level, so the raised bed wall doesn't tend to buckle. If the ground is too uneven for this, just cut and rejoin the mesh to accommodate the ups and downs. Also where there is a pucker at the top, just use pliers and hammer to make a fold in the mesh.

For the lip of the raised bed, make a stiff mix with fibre and hand plaster a nice thick line over the edge (after the first coat has dried).

In a long straight line where the mesh wants to fall over, put temporary wooden stakes in to hold it, or you can wrap the strip around e.g. a piece of pipe or dowel to make a loop in it, big enough to stabilise the strip on the ground. Or hammer the dowel into the ground and leave it—it will rot but this won't matter. For extra strength, especially in straight lines, lace 6mm rod to the top of the strip. Or, ideally, fold the rod into the strip.

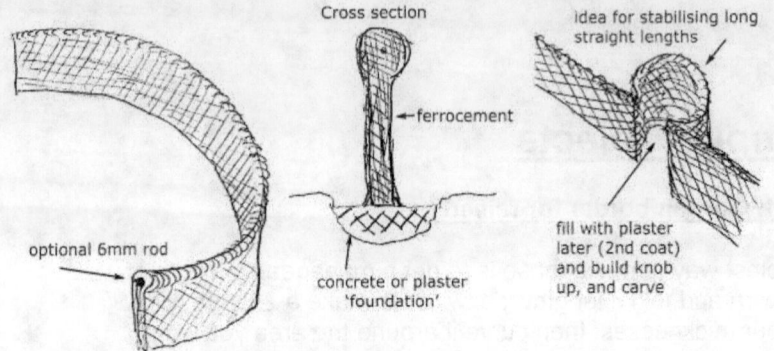

Cross section

idea for stabilising long
straight lengths

←ferrocement

optional 6mm rod

fill with plaster
later (2nd coat)
and build knob
up, and carve

concrete or plaster
'foundation'

FIGURE 34 Raised bed borders

Update 2012: See the photos at the end of this book, for my latest raised bed. Needing a high wall (about 350 mm) I folded the chickenwire only once, and with plenty of fibre in the mix it plastered OK with only one coat. Also see the method I tried using 6mm 1200 mm lengths I had been given. I have also made a raided bed for Anna at her new house in Gisborne, folding the netting into three instead of four or two. This gave a height that was perhaps the best for all-round raisded bed gardening.

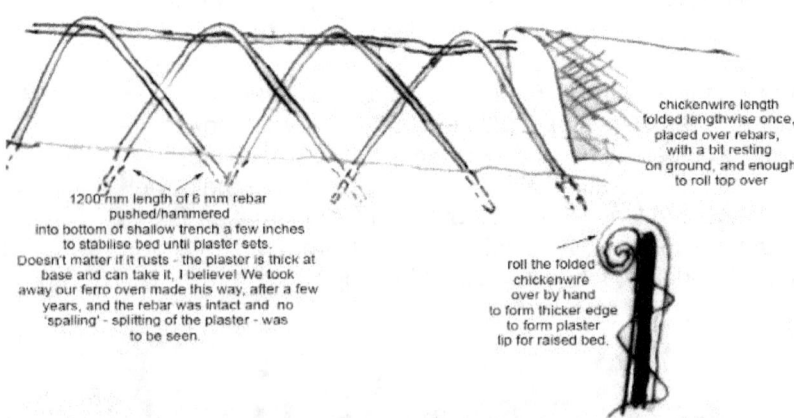

chickenwire length
folded lengthwise once,
placed over rebars,
with a bit resting
on ground, and enough
to roll top over

1200 mm length of 6 mm rebar
pushed/hammered
into bottom of shallow trench a few inches
to stabilise bed until plaster sets.
Doesn't matter if it rusts - the plaster is thick at
base and can take it, I believe! We took
away our ferro oven made this way, after a few
years, and the rebar was intact and no
'spalling' - splitting of the plaster - was
to be seen.

roll the folded
chickenwire
over by hand
to form thicker edge
to form plaster
lip for raised bed.

FIGURE 34a A more recent design I used.

Project 2. Angel/Gnome

This is an example of a small sculpture (say 600mm high), which may only need an armature of thick lacing wire. One good way to make the basic armature is to make rings of wire as horizontal circumferences for several levels of the sculpture. Then tie (or loop) vertical wires between them, to make the body of the angel. Then add loops for the arms and wings, lacing wire around them to form a crisscrossed framework. To adjust the shape, you can bend little loops into the wires with pliers to shorten them.

Then wrap the armature in pieces of chicken wire and lace them on, either reaching in to stitch or using the semicircular needle.

To plaster, use thick coats and leave plenty of 'meat' for carving the final form. Remember, the armature is just to form the 'skeleton', so it need not have the exact shape of the final angel, let alone the details, e.g., of fingers, hair and face.

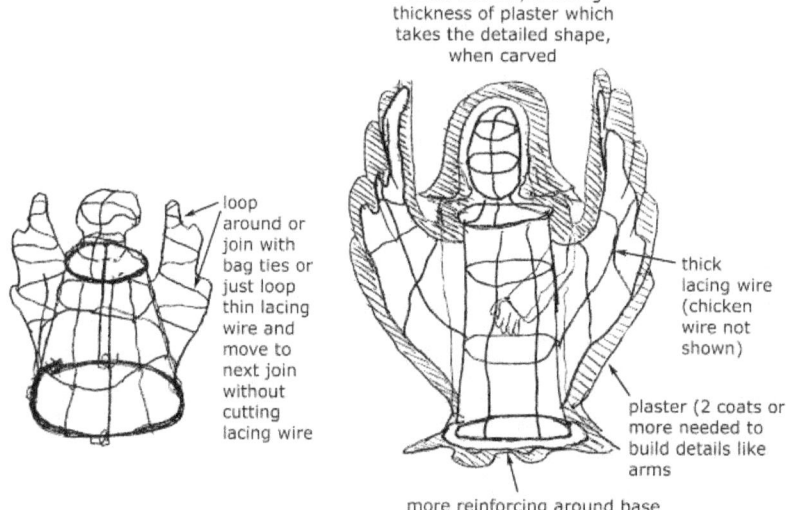

Cross-section, showing thickness of plaster which takes the detailed shape, when carved

loop around or join with bag ties or just loop thin lacing wire and move to next join without cutting lacing wire

thick lacing wire (chicken wire not shown)

plaster (2 coats or more needed to build details like arms

more reinforcing around base

FIGURE 35 Angel

Project 3. Garden Pot

This can be made almost as a basket is woven. To be worth doing in ferrocement, it will be quite a big pot, say a metre wide and 400mm deep. Or maybe a cauldron shape, around 800mm diameter. Let's do the 1m by 400mm—I made one for the centre of our parterre and it was great.

The first thing is to make the vertical ribs, which also form the floor of the pot. These are best joined on the middle of the bottom with a ring of the same 6mm rod.

Make a half-profile of the pot for bending the rods to. Either drive thin stakes into the ground in the desired shape, or say 75mm nails into plywood; or cut a permanent template out of thick ply or particle board, with a loop of wire or similar to put the end of the rod into when you bend it. Make the profile smaller and higher by about 20%, to allow for the springing back of the bent rod. When you cut the rod to length for the vertical ribs, allow for about 50mm extra at the top so you can bend the vertical ribs around the horizontal rib at the lip of the pot.

Then make the horizontal ribs, loops of 6mm rod, say two for the bottom and two for the sides and two for the neck and lip of the pot— a total of six, taking the diameter by laying out two of the finished vertical ribs and measuring the diameter at the six positions. Tie the horizontal loops to the vertical (help in holding them up is handy).

Then, to fill in the shape, bend thick lacing wire around the ribs between the 6mm ones, until you have a grid of say 100 mm squares. Then chicken wire in the usual way. If you want a garden pot, with soil, put in a few holes with plastic pipe or similar plugs to stop them being plastered over. These are to let the pot drain. If you want to use the pot as a pond, leave the holes out. They can be plastered over later to make a pond, of course. For plastering, sit the framework on a few bricks or similar, so that you can plaster underneath as well as inside. This will leave patches that need a bit of plaster on the bottom after you turn the pot over when it is cured.

It is possible to do it all (apart from the patching) in one coat, but two may be better for smoothness of the shape and full coverage. And for carving. It is easy to carve repeating patterns in the lip to make a classical look—as long as you put plenty on so you don't carve through to the mesh.

Although the finished pot will be heavy, it is strong enough to be rolled and otherwise manhandled. See Moving ferrocement objects.

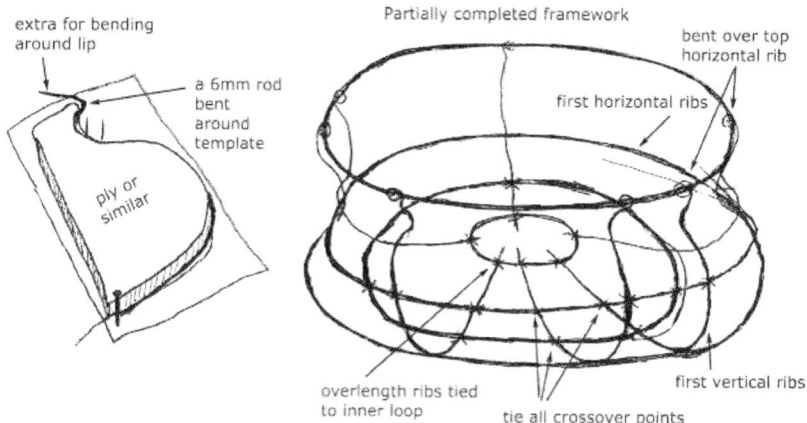

FIGURE 37 Pot framework

Project 4. A 2.5 Metre Dome

To give full detailed instructions for a 2.5metre dome with lighting and windows and door is beyond the scope of this book, but I will try a general instruction for it, which if you have grasped all the techniques above, should be enough for you to make it without much trouble. Certainly, you will find it easier than we did the first time!

Poured Concrete Foundation

This is easily laid in a boxing made of e.g. strips of thin ply say 100mm wide, taped/glued together to form a circle say 2.7metres diameter, allowing a 100mm edge beyond the wall. Use the usual 665 mesh reinforcing, and starters 600mm apart in a circle 2.5metres diameter.

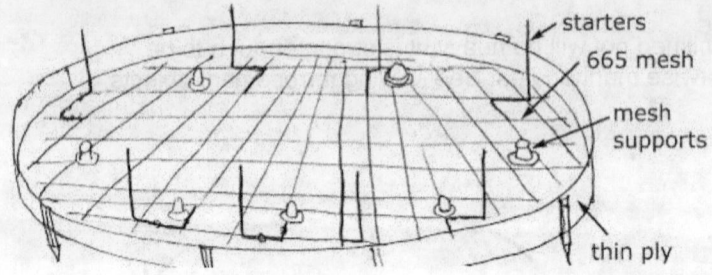

FIGURE 37 Dome foundation

Wall

This is made by bending sheets of 665 reinforcing mesh around into a cylinder 2.5metres wide, tying it securely to the outside of the starters. For a full 2.5 metre height you will need to cut another strip and overlap and tie it on top of the first cylinder, inside it so the wall steps slightly inwards rather than outwards. For the gutter, cut the mesh at the top so one mesh level can be bent over at right angles. Bend a 6mm rod around the gutter perimeter and tie. Tie several wires across the diameter, measuring as you go to pull the top of wall into a perfect circle (it will be a bit smaller than the 2.5metres because of the overlapping mesh). Then bend a few more 6mm rods around under the gutter for rigidity, tying well.

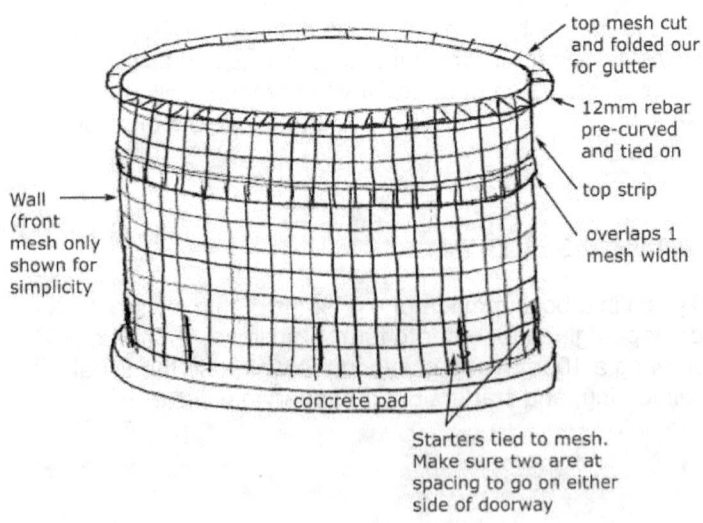

66

FIGURE 38 Dome wall frame

Door and window openings

Mark and cut to nearest mesh line. Bend 12mm reinforcing to make
frames for openings. Make 225mm wide strips of chicken wire
folded lengthwise twice, and wrap around inside of openings,
preferably with 6mm rod bent around also and laced to the chicken
wire, to strengthen it. Remember to make openings about 70mm
wider and higher than the frames of the doors and windows. (See
section on doors and windows).

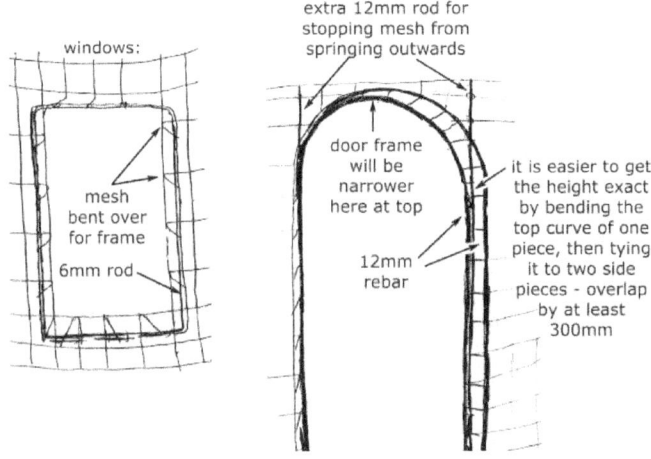

FIGURE 39 Door frame

Roof

First make the vertical ribs, eight or so, of 12mm rod. See drawing
for profile. Tie the ribs to the inside of the wall. Make horizontal
rings to fit over the top of the ribs, and tie them on. Then bend 6mm
horizontal ribs around the verticals, around 150mm apart, tying
each junction with a bag tie. Then bend some more verticals, 6mm
this time, to go between the 12mm ones, and make the mesh of the
roof dome about a 150mm square at the bottom, (and
automatically, smaller at the top).

67

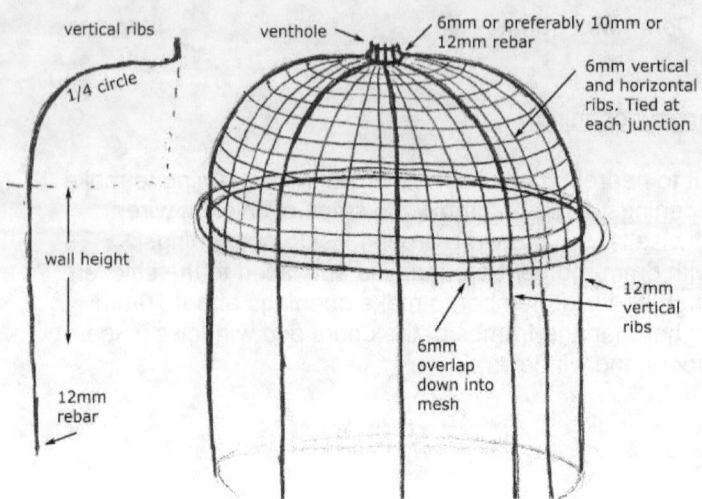

Labels in figure:
vertical ribs
1/4 circle
wall height
12mm rebar
venthole
6mm or preferably 10mm or 12mm rebar
6mm vertical and horizontal ribs. Tied at each junction
12mm vertical ribs
6mm overlap down into mesh

FIGURE 40 Dome roof frame

Chicken wire

The walls are easy; the roof needs to have the chicken wire strips cut into long triangles, so they can be laid vertically down the roof. Do this by cutting the strips off the roll, half as many as you will need to go around the 7.8metre circumference, allowing a 50mm overlap at the bottom. I make it 9.17 strips, say 10, times 4 for the four layers = 40, divided by two since you will get 2 triangular strips per rectangular. So, 20 strips 1.964 m plus 100 mm for overlap = say 2metres.

Lay all 20 rectangles on top of each other neatly, then cut along the diagonal using the angle grinder, kneeling on a long board which acts as a guide as well as holding the mesh down. Then lace the triangles starting from the top, 2 on each side of the roof. This is much easier with two people! The top where there is a vent hole, and the bottom where the guttering ring is, are tidied up when you lace chicken wire onto the vent and gutter.

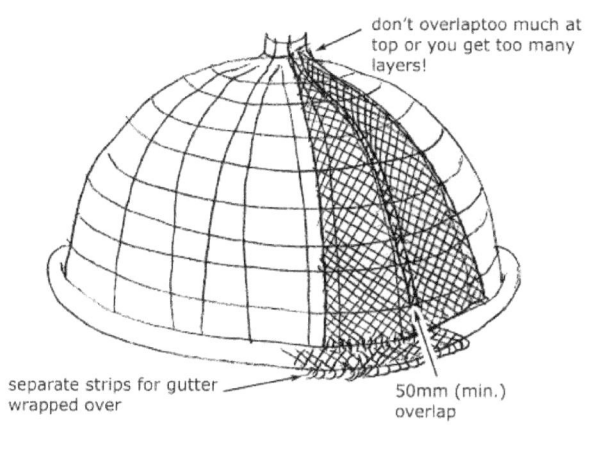

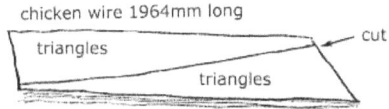

FIGURE 41 Dome chicken wire going on roof

Plastering

Once all is neatly laced, with no puckers or loose wires or protruding reinforcing, and all conduits, hole plugs, etc, are in place, you are ready to plaster. Start with the roof; the framework is strong enough to support the plastered roof. Plaster inside only when the outside is done and plaster is extruding out the mesh on the inside, and when it has begun to go off, and is not too runny. Use the absolute minimum of trowel strokes, or you will get water coming out and plaster falling off the ceiling! The second coat is for the final smoothing. Scratch all over so the second coat will key, otherwise you will have to paint the whole surface with adhesive.

Cap for vent

Use the instructions for pots—or even simpler, sandcast a 'hat' using a circle of ply or a big metal lid etc, to scoop a semicircular depression in wet sand. Then carefully pat a thick fibrous plaster mix onto the sand wall, and a bit of a lip over the edge of the hole, and wait for it to dry.

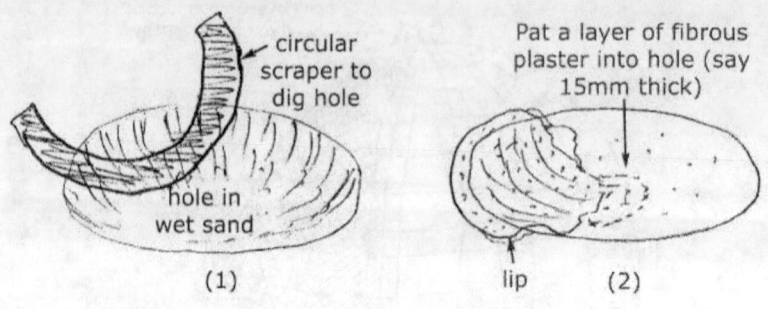

circular
scraper to
dig hole

hole in
wet sand

(1)

Pat a layer of fibrous
plaster into hole (say
15mm thick)

lip   (2)

FIGURE 42 Sandmoulding vent cap

Congratulations! You have now made the shell of a permanent
dome. Second and carving coats to follow. And paint.

Insulation

If you want insulation, my preferred method is to glue strips of
20mm polystyrene onto the inside walls and ceiling, two layers
preferably, and plaster over it. Without this, you will find the dome
gets cold in winter and loses heat quickly. Ferrocement is not an
insulator!

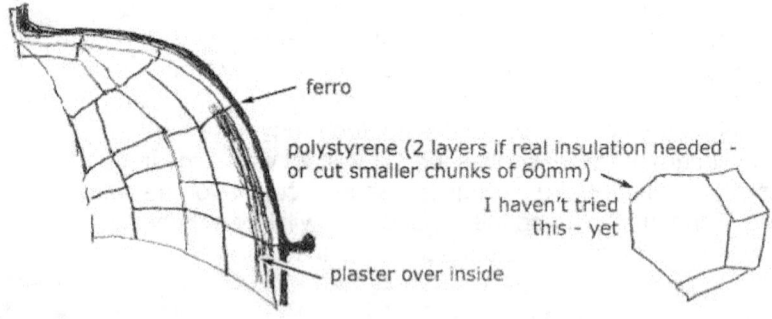

ferro

polystyrene (2 layers if real insulation needed -
or cut smaller chunks of 60mm)

I haven't tried
this - yet

plaster over inside

FIGURE 43 Polystyrene insulation

Pillars

The best method is I think to hang 100mm spirals of thick lacing wire, two pushed together to make a double spiral, from the gutter to the ground. Then to solid plaster using a fibrous mix with minimum water (usually two or three coats to build up the full 200-odd mm thickness of the pillars.

For branches off the pillars, just freehand place and shape thick lines of plaster–again, two coats or so.

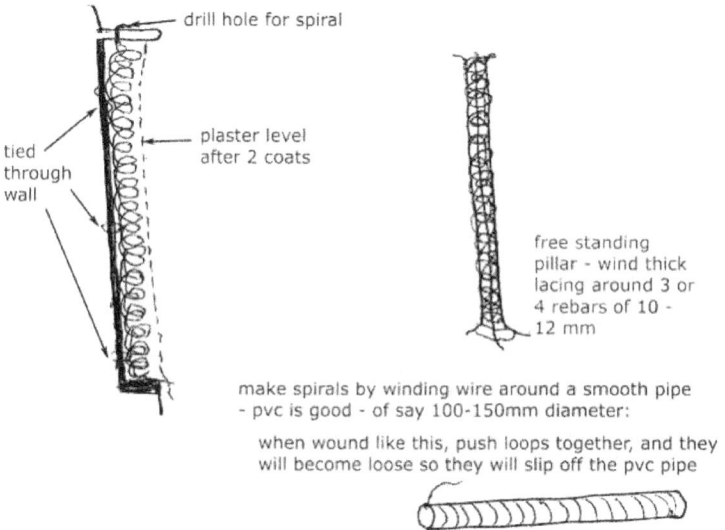

drill hole for spiral

plaster level
after 2 coats

tied
through
wall

free standing
pillar - wind thick
lacing around 3 or
4 rebars of 10 -
12 mm

make spirals by winding wire around a smooth pipe
- pvc is good - of say 100-150mm diameter:

when wound like this, push loops together, and they
will become loose so they will slip off the pvc pipe

FIGURE 44 Pillars

Watertanks

I have never got round to making one of these. Companies like Absolute Concrete here in Kaiwaka make a whole business of manufacturing a range of them, at such good prices (when you take your own labour into account), that it doesn't seem worthwhile to go DIY. A big tank has tons of concrete in it, and would take almost as much time to do in ferrocement as a dome, and probably needs thicker walls and more reinforcing, as it is holding 20 or more tons of water. Get a quote at least, before embarking on building a concrete tank. The manufacturer also transports and sites the tank, and this is a big factor too. I would only make a tank if it was an integral part of a building, and had to be an odd shape. Then, ferro would be ideal!

One idea I do want to try is to put a ferro 'rampart' around the top of our water tank, with a spiral stairway up to it, and perhaps a ferro roof on pillars, for an elevated gazebo. The tank has the new-style flat roof which makes this an easier proposition...

Ferrocement bushbath

I made one of these easily, using corrugated iron hammered lengthwise into the ground in a shape to fit around an old iron bath. I left a gap at one end (no need for a door) put a length of iron down for a grate, and at the far end made a hole in the iron and stuck a steel pipe chimney into it at an angle. Then I put pink fibreglass `Batts' around the iron, tying it on with wire, then wrapped chicken wire once around the batts and plastered it with plenty of fibre, plastering up to the rim of the bath, and making some decorative pillars. This gave us a well- insulated good-looking bath that heats up fast.

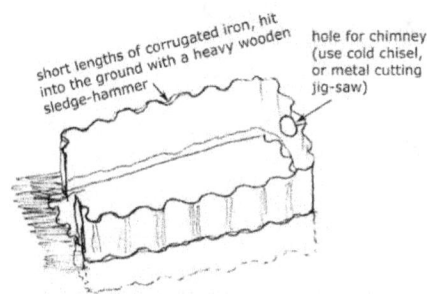

short lengths of corrugated iron, hit into the ground with a heavy wooden sledge-hammer

hole for chimney (use cold chisel, or metal cutting jig-saw)

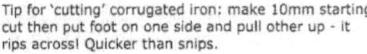

Tip for `cutting' corrugated iron: make 10mm starting cut then put foot on one side and pull other up - it rips across! Quicker than snips.

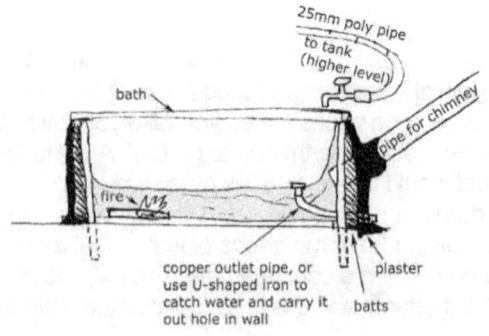

25mm poly pipe to tank (higher level)

bath

pipe for chimney

fire

copper outlet pipe, or use U-shaped iron to catch water and carry it out hole in wall

plaster

batts

72

7 Photos and updates to the Eutopia story

Foundations. Note power conduits, drainpipes (lagged) and mesh held off ground

Pouring foundations. We tied the starters to the mesh for accurate placement. We over-specified the cloister floor at 200 mm thick - 100 mm would have been fine. (I was worried about getting a permit, which had this effect!)

Doors - experimental, plywood. Jigsawed out, routed the rebates for glass.
Best to use at least the 17.5 mm thickness - I used 12mm and they warped
a bit, also not enough clearance for the glass.
I bolted wood to the pillars, then plastered gaps. In later years the
untreated ply did start to disintegrate and needed patching up. Worth
always treating the ply with copper-based preservative after cutting out! Or
use treated ply - though I don't like working treated timber...

Earth pizza oven on ferro base / firewood store. I painted the ferro base later. In the end a local man bought it as we didn't use it enough at the cafe. He successfully hauled it up with a hiab truck and transplanted it onto his farm.

Framework of bridge. Fountain covered - plaster does splash!

Help is nice. Our sons. Note needlenose pliers - good for tightening chicken mesh

Framework for low walls of fountain - note no rebar, just rolled up mesh for thicker edge.

Anna with rose. Framework nearly done! Note lip of fountain - extra chickenwire.

Raewyn roped in! Note stitching needle going through.

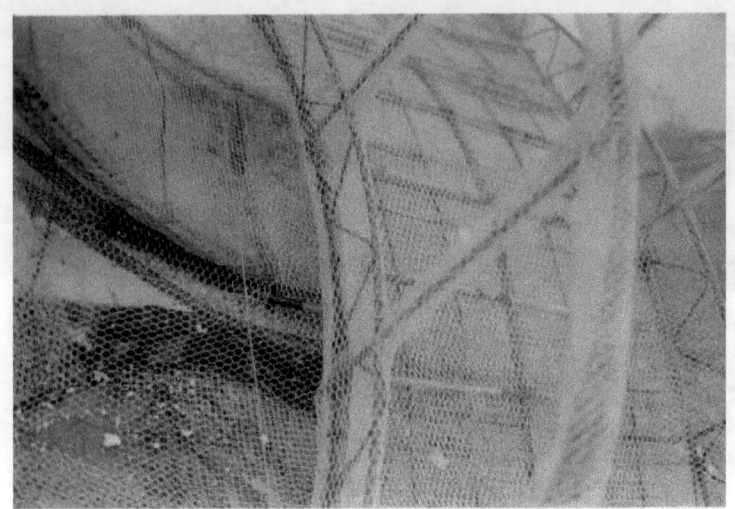

Framework of bridge. Note zigzagged 6mm rebar

Lacing takes time! Anna being patient

This stage can drive you mad! Help is good, to hold up mesh pieces for lacing.

Freeform framework of arches - note 6mm rebar bent over to form mesh. Also note bent-over ends. the pipe bender does these nicely with a bit of practice.

Monkey business. Rare camera-shy ape of the ferro forests... Note strength of framework already.

The author, resting... this takes practice!

Carved tabletop - ordinary ferro plastered, carved with scraper and sandpapered well. Bronze finish is antique gold acrylic paint, overpainted when dry with dark green acrylic, and the excess of the green wiped off, leaving green in the hollows and clean bronze in the relief parts, then varnished.

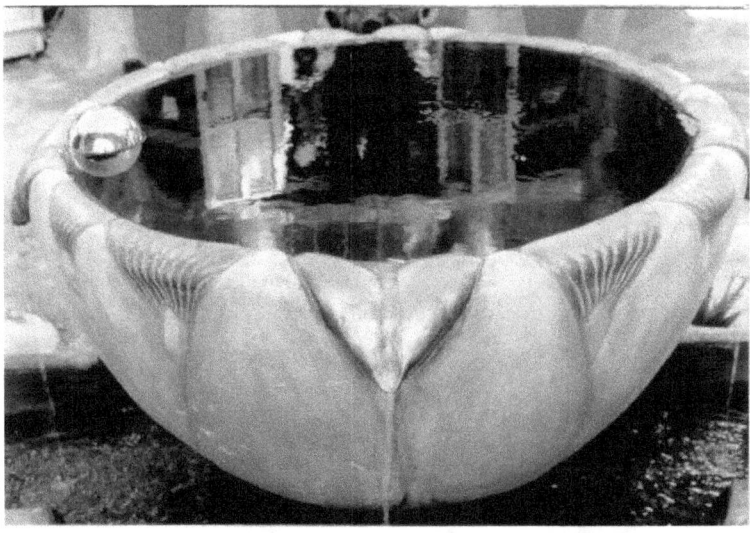

Fountain, painted. The tips of the pourers I made with epoxy for accuracy and toughness. Not really necessary though. Paint: Blue acrylic exterior, brushed over white undercoat, varnished with clear exterior varnish (the latter not strictly necessary I think - since then I have repainted with just acrylic).

Carved pedestal. Note pillar ready for second coat. Pipe to feed rainwater to sculptural gnome.

Framework. Note bender I am using, and Daniel with the bagtie-twisting hook in drill. Lots of glare out in the open, hence the heavy sunglasses!

Plastering moongate - I'm wearing nice 'stretch' raingear including leggings - the girls aren't so well-prepared!

Good plastering weather - overcast but not raining. Note heavy gloves for first coat, pushing plaster into mesh.

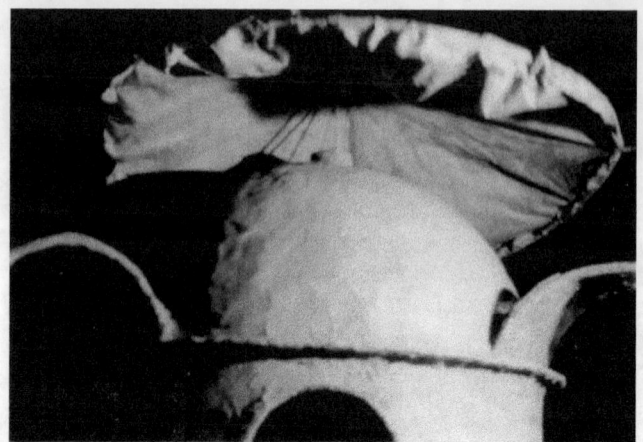

Wet weather, midnight plastering before opening when the mayor was coming...

As our son said before he went to bed at 2 AM or thereabouts, 'This is madness!'. At least we had good light - 500 watt floodlight. The drips of rain from edge of tarpaulin etc caused runs in the plaster which you can still see on that dome today...

Macrocarpa slab door, plate steel hinges cut out by welder. Cost $200, but I thought worth it!

Crane lifted this cupola onto kitchen easily. Anna was scared though!

Anna where it all began: Shackleton road fairy garden and the first dome.

Cafe moongate 2009 - still good paint etc. Gold acrylic lasted well too, on roof. 4-5 years so far

An early garden pot I made in Shackleton Road before heading North to Kaiwaka.

Cafe courtyard 2009. Fountain on narrow 'stalk' of ferro still stands strong 8-odd years on. original paint apart from added gold on rim.

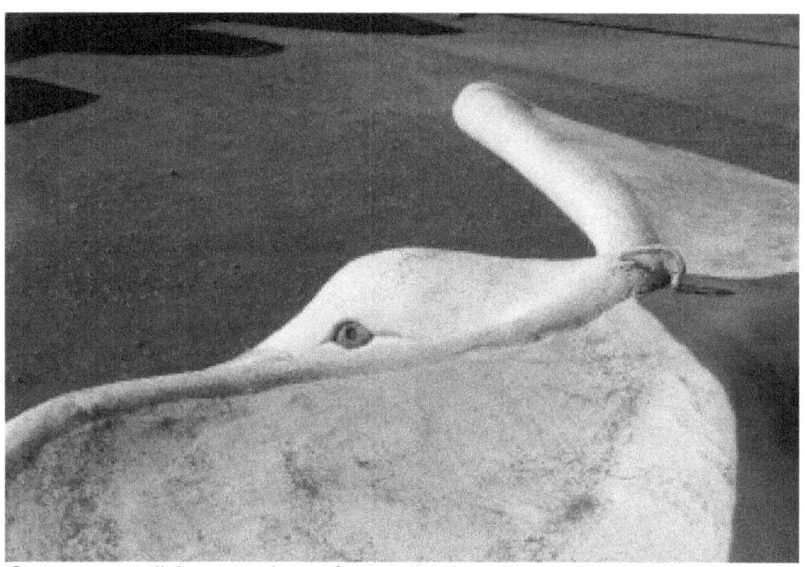
Some orange lichen on wings of seagull; otherwise white acrylic as new, 4-5 years on. No cracks on wings.

Kids love the new prow! Note the fineness of structure - still VERY strong! Passers-by see the cafe more easily now, and stop to have an organic espresso or juice and take lots of photos. And buy the books and jewellery! it is great to have a unique place - worth all the work.

An experimental garden shed on our land using 3 by 2 inch wooden frame, cement plaster over stapled-on fibreglass cloth mesh for roof and plaster of Paris over the same for walls. Cloth sagged on roof - needs to be stretched and stapled tightly, or else use plaster of Paris in thin first coat to give rigidity before coating with more.

2009: the author and wife Raewyn in experimental 'bamcemboo' dining room.
Bamboo framework, quickly assembled and held together with packaging tape, then wound with double spiral of low-tensile wire about as thick as fencing wire, instead of ferrocement 12 mm rebar for main framework. No guarantees, but promising so far! Plastered as for ferro, but fiberglass mesh in place of chickenwire, and polystyrene tied onto frame for insulation. A book on this development may soon be warranted. One planned use is for large tunnel house (hothouse) frames, using bent-over bamboo poles lashed together and plastered, forming 4 metre semicircular frames for the polythene. Watch this space!

The work has gone on in short bursts of bank-funded activity, but the ship's prow in 2008 was the only major for several years, due to many other activities of dubious   profitability, such as the e-press digital publishing printing and binding. There has been the writing and rewriting of the five volume epic Apples of Aeden, (see www.applesofaeden.com), the Passport to Eutopia (see mapofeutopia.com for a free pdf in colour of the 64 page book of the process philosophy and ideals behind Eutopia and Dreamspace, the IdeaTree series begun with How to be Creative: A Passport to Creativity) and the gradual improvement of our improvised dwelling on our ten acres by the kaipara harbour. Not to mention the years of running Cafe Eutopia before selling the business, in order to focus on the continuing building of Eutopia, and my writing.

Thanks for buying this book, and may your creations be joyfully completed and bring inspiration to others for many years! Though we have slowed in our project, it is a constant comfort to know that those parts that are done will outlast us, and inspire long after the initial cost is forgotten.

Peter Harris 22nd October 2009

2012 Update to the story.

A selection of photos taken in April 2012:

Signwriting which replaced the chairs (as seen on cover) has held up well - red roof paint stencilled on using large printouts of the letters.

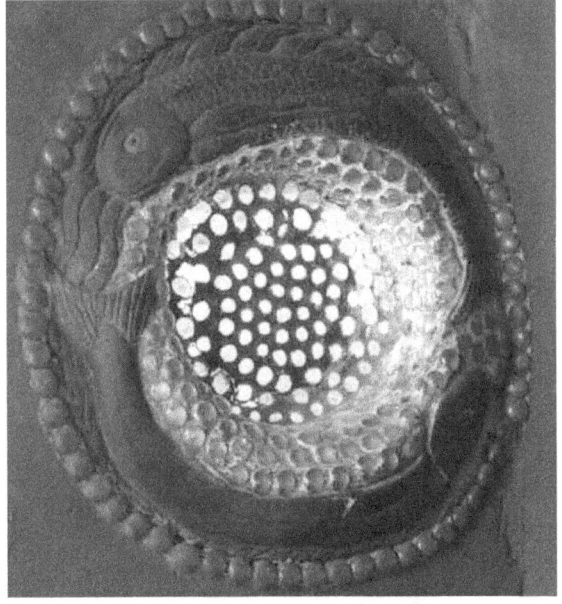

The fish were carved - about 10 years ago, this is the original paintwork - as described from a second/third coat put on thick in rough shape desired, then scrape-carved with tungsten scraper.

91

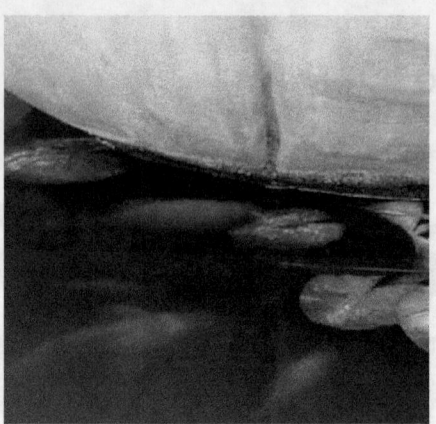

The goldfish have thrived in the fountain. Some have been there for years, and grown quite big - about 6 inches. And had babies. The biggest has nibbled fingers, and the cafe girls named it Pig.

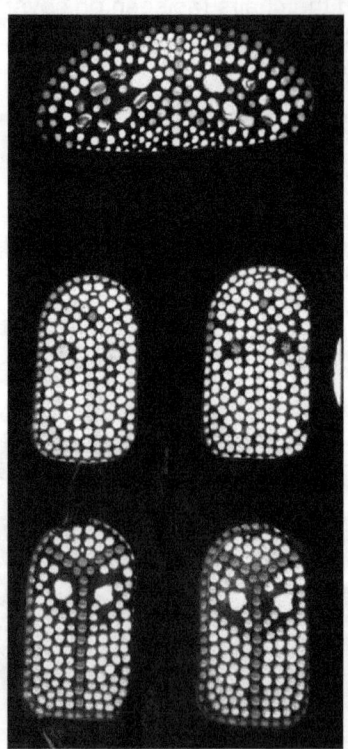

Stained glass in the Blue Dome. Window holes cut out of two macrocarpa slabs.

Ceiling of Blue Dome. Note the finish - dark blue acrylic rubbed with a sponge with a dab of antique gold acrylic on it. Light is made as described above, and would be brighter-coloured if I had used a clear glass globe instead of frosted plastic one.Convex mirror works well in a dome...

The girl angel twelve years on. She's been repainted a bit and recarved.

A new raised bed being made with only two thicknesses, i.e., one length folded lengthwise. 6mm rod reinforcing, bent to L shape, pushed into ground to support it, and one 6mm along the top, woven in and out of the L shaped pieces. Rebar not even tied, just the chickenwire laced. It is rolled over at the top.

The whole raised bed, made for Raewyn, at our Story Ark across from Eutopia, where I now live and work, making books as New Leaf Network, the Book Wizard, and inventing and writing...

The plaster is one coat only, roughish and wobbly lines, but it is very strong! And it only took about a day all up. It is about 14 metres long and took two and a half 40 kg bags of cement. I stapled the ends of the rebar in the short lengths to the wooden fence. Should serve as a retaining wall against the slope from neighbouring property. Stained by painting a solution of sulphate of iron onto the fresh plaster. Nice colour, though it doesn't soak in as deeply as I had hoped, and does scratch off if scraped, eg with a spade.

Raewyn is now studying psychology and looking after the mentally handicapped, having finally decided she can't be a wizard's 'business' partner. (We are still married though! The pursuit of the ideal has its costs, but if we honour each other it all turns out OK in the end. Or at least, that's my story! Or to quote the Indian saying, 'It will be all right in the end...And if it isn't all right, it isn't the end.'

Just today when taking the photos, a man I didn't recognise greeted me warmly and said if we can rave as we work he would gladly volunteer on my next 'madness' - I had only to ask! He cherished the time we had 12 years ago when he helped for a bit in the first one...

So we never know the effect we are having by just going out and creating according to our vision. Ferrocement follies to some, to others the seemingly ephemeral visions of beauty captured in permanent concrete for those who come after us... There are twelve-year-olds who smiled at the Eutopia fountain as babies, who now come back to see it, and perhaps one day they will return again when they have babies of their own, and they too will smile with joy to see the fountain.

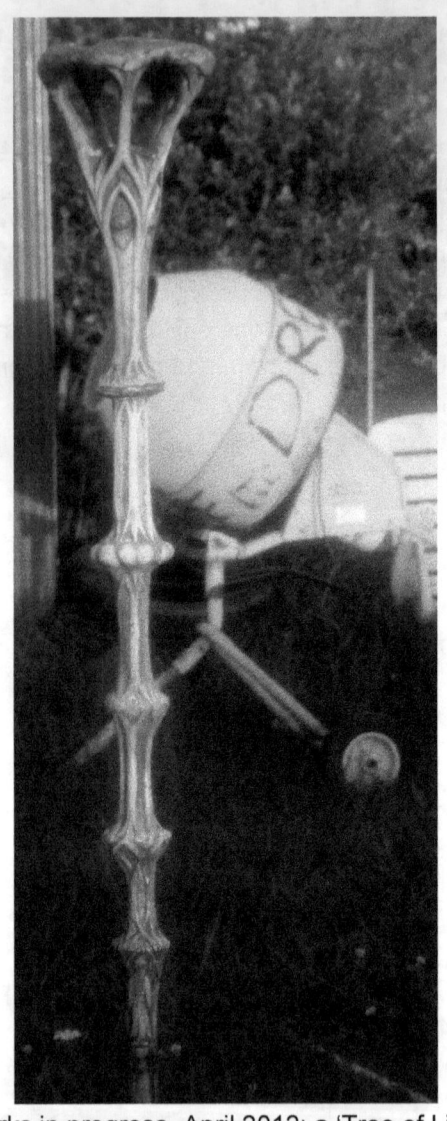

One of the works in progress, April 2012: a 'Tree of Life' staff, carved from plaster of Paris on a framework of five 6mm rebars. Heavy? Yes, but I will make a mould and cast some lighter ones... Could cast in cement too for outdoor fixture... Note the old faithful cement plaster mixer in background.

The Restoration of the Balance

P.J.Harris, Café Eutopia, Kaiwaka

First sketches June/July 09 for the sculpture to top the Dome of the Courtyard, Norman and the Lady replacing the Heartstone into the Tree of Life as per Volume Three of 'The Isle of Eden'

Another of the grander visions: concept drawing for a possible sculptured dome over the courtyard... Based on my 'Apples of Aeden' epic. As is the Tree of Life staff.

Happy dreaming and building, in a happy balance! And let me know what you get up to!

8 Further Sources

Internet

There is quite a lot on the internet, one way or another, though the fashion for ferrocement boats has waned. There is lots on other related media like fibrous cement, sprayed concrete over rubber moulds, etc.
I am trying to collect any articles that come up on the Web, at my new blogsite **www.fantasticferrocement.com**
Try 'ferrocement' in a search engine. There are quite a few pictures of ferrocement houses etc. Also try http://www.ferrocement.net – Paul Sarnstrom has reviewed this book there (favourably!)

Other Books

Try your library, but the books are likely to be on ferrocement boats, a far more exacting pursuit, and are likely to be oldish, so some things could have been improved on. E.g., I read one book in the 60's that recommended a framework of pipe for ferrocement boats—but the pipe corrodes from the inside, so is not a good idea!

University Engineering and Architecture libraries will have some books and references to ferrocement, too.

People

Plasterers know a lot that applies to ferrocement, so if you can employ one for a day, or work for one just to learn more, this would be great.

Yellow Pages under:

Cement suppliers, especially plasterer's suppliers, will have products and contacts and helpful advice.

Paint suppliers can advise on the various concrete paints – there are so many we don't want to say what is best. But we have found Resene to be good – for the shiny hard finish we needed for our kitchen floors and benches, their Aquapoxy was great – no smell as it is waterthinned. And for our gold dome roofs, their metallic range was good.

A friendly concrete tank maker could be helpful—their plasterers could be interested in a little overtime helping with your project, at least the plastering stage.

The Author:

peter@eutopia.co.nz, Box 37, Kaiwaka, Northland, New Zealand. Ph 09 431 2178. See our website for some more pictures: www.eutopia.co.nz Or - this is a wonderful development with the Web - if you go to Google Images and search under 'Cafe Euopia' you will see a whole lot of photos from the many good folk who have taken photos and uploaded them.

We are not set up as ferrocement builders or consultants as such but would be interested in what you are doing and may be able to help with your project. And any photos and tips welcome too! We can add them to future editions...

More things by Peter Harris

If you liked this book you may be interested in my philosophy and fiction which is of course steeped in it.
Non-fiction:
- The IdeaTree series, expounding the five zones of the Tree of Life or 'IdeaTree' diagram of all process and development.
So far the second book is out: 'How to be Creative - a Passport to Creativity'.
- How to be a Wizard - How life is magical and we are too
Fiction:
- The fantasy epic *Apples of Aeden*.
- T*he Sword of the Fifth Element*, set in the Dark Ages in Cornwall and the world of Aeden.
- *The Nautilus Project*
For all my latest ebooks including the ebook of *Fantastic Ferrocement*:
www.smashwords.com/profile/view/wizardofeutopia
or amazon.com/author/peterharriswizardofeutopia
For YouTube videos see my channel:
www.youtube.com/user/wizardofeutopia?feature=guide
For my general blog which has some free posters and ebooks, and reflections of philosophy, architecture, art, book design and writing, the new economy and civilisation, go to:
www.wizardgifts.wordpress.com

Also from there you can go to all my other blogs, such as
www.carvedbooks.com,

where you can see coloured photos of an invention I call 'carved edges' on books:

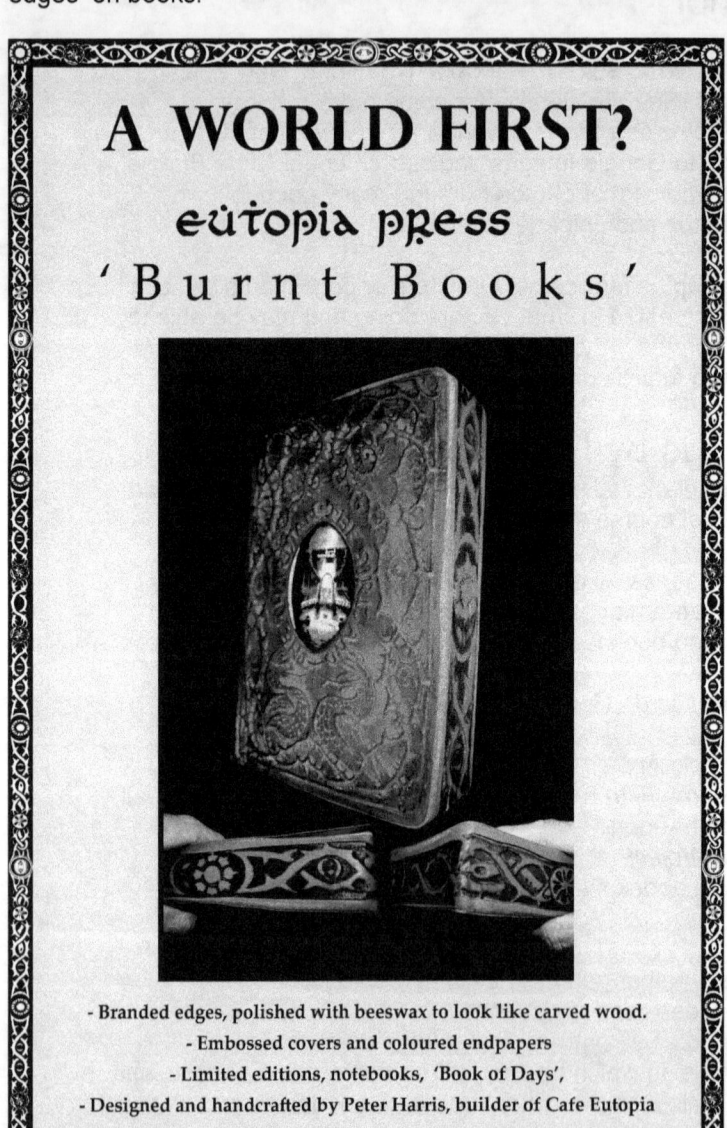

A WORLD FIRST?

eutopia press
' B u r n t   B o o k s '

- Branded edges, polished with beeswax to look like carved wood.
- Embossed covers and coloured endpapers
- Limited editions, notebooks, 'Book of Days',
- Designed and handcrafted by Peter Harris, builder of Cafe Eutopia

www.ingramcontent.com/pod-product-compliance
Lightning Source LLC
Chambersburg PA
CBHW061514180526
45171CB00001B/171